The New Space Race

China vs. the United States

Erik Seedhouse

The New Space Race

China vs. the United States

Published in association with
Praxis Publishing
Chichester, UK

Dr Erik Seedhouse, M.Med.Sc., Ph.D., FBIS
Milton
Ontario
Canada

SPRINGER–PRAXIS BOOKS IN SPACE EXPLORATION
SUBJECT *ADVISORY EDITOR*: John Mason, M.B.E., B.Sc., M.Sc., Ph.D.

ISBN 978-1-4419-0879-7 Springer Berlin Heidelberg New York

Springer is a part of Springer Science + Business Media (*springer.com*)

Library of Congress Control Number: 2009936076

Cover design: Jim Wilkie
Project copy editor: Christine Cressy
Typesetting: BookEns, Royston, Herts., UK

Printed in Germany on acid-free paper

Contents

Preface

OUTLINE OF THE CHAPTERS

This book examines the civil and military space programs as the two sources of competition in the impending space race between China and the US. The book is organized into four sections. Section I characterizes China's long march into space and provides an insight into the space policies of the US and China. Chapter 1 focuses in particular on the impetus behind China's nascent space program before examining the history that led to China becoming a tier-one spacefaring nation. Chapter 2 provides an outline of Washington's and Beijing's civil and military space policies, focusing on policy goals and objectives.

Following the political groundwork, Section II provides an insight into how an arms race in space may evolve. Chapter 3 shows how the space warfare doctrine of the US is designed to achieve full-spectrum dominance, whereas the doctrine of China is to develop a preemptive strategy with the goal of defeating the US asymmetrically. Chapter 4 assesses the space weapon capabilities of the two countries and how these weapon systems might be employed in a future conflict. Next, Chapter 5 describes the concept of *space dominance* and how the US plans to ensure space superiority by seizing hold of the future of war. This chapter then assesses the asymmetric advantage and vulnerability that the US enjoys and suggests ways in which China may react by developing counterspace capabilities. The final part of Chapter 5 describes two hypothetical scenarios in which China could win and lose a space war with the US. At the conclusion of Section II, it is posited that although the extent of Beijing's pursuit of space weapon technology is uncertain, a new arms race in space is not unthinkable. Such an aggressive stance is proposed because China's statements purporting to use space for peaceful purposes are nothing more than empty rhetoric designed to disguise its real intentions to deploy its own space weapons.

The focus of Section III is the second component of a future space race. Whereas the first space race was characterized by the Soviet Union and the US racing to the

Moon, the objective of the new space race is nothing less than leadership in space. Chapter 6 provides an insight into China's and the US's space exploration programs. Whereas NASA's Vision for Space Exploration is funded to the tune of several billion dollars a year, China's entire annual budget is barely two billion, yet both programs have the same goal. This chapter explains how China is able to achieve so much with so little. Chapter 7 reveals details of China's space technology and how it compares with NASA's hardware designed to return astronauts to the Moon by 2020. This chapter also explains how China may be able to compete with NASA by skipping generations of technologies by buying and absorbing foreign expertise. Chapter 8 focuses on the question of whether China, a neophyte in the world of manned spaceflight, can hope to compete with the US, which routinely chalks up more manned spaceflight experience in a week than the cumulative total of all China's missions.

Finally, Section IV analyzes the factors described in the previous sections and asks how a space race may be avoided. Chapter 9 considers the case for and against collaboration with China and suggests that any attempt at cooperation is doomed to failure in light of the strong anti-China undercurrent in present American conservative politics. Chapter 10 concludes that Beijing cannot be trusted with regards to spaceflight or geostrategic intentions and, given the prominent challenge represented by China, the strategic landscape of the new space era is about to be forever altered by a contest in space.

THEMES AND OVERALL ARGUMENT

This book argues that there is compelling evidence for an impending space race between China and the US. Driven by ambitions to place astronauts on the Moon and driven by fears about national security, the new space race will undoubtedly be fought on two fronts, the first being in the manned spaceflight arena and the second in the strategic dimension. To that end, Beijing has read the playbook of NASA's space program and has decided to pursue manned spaceflight for many of the reasons that the Americans do, such as enhancing international prestige and advancing science and technology. China has also taken note of the US's effort to militarize space and to establish unilateral hegemony and its avowed intention to ensure unrivaled superiority in space, as evidenced by its provocative demonstration of ballistic efficiency when destroying one of its own derelict satellites in January, 2007. Additionally, China's anti-satellite (ASAT) test not only signaled that China had become the challenger to the US, but that space had become the new territory for military competition.

On October 15th, 2003, China became the third nation to independently launch an astronaut into Earth orbit, four decades after the Soviet Union and the US first sent men into space. While the event that matched the feats of the Soviet Union and the US was noted by many as a milestone in human history, China's first manned spaceflight may, in due course, be remembered as the event that launched a new space race. But, whereas the first space race was characterized by the goal of a "flags and footprints" mission to Earth's closest neighbor, the prize in the imminent Sino–US competition is nothing less than total military domination of the space environment.

The first space race began on October 4th, 1957, when the Soviet Union launched *Sputnik I*, the world's first artificial satellite, a feat that forced the US to accelerate its fledgling space program. On January 31st, 1958, the US launched *Explorer I* – an event signaling the beginning of a decades-long competition in low Earth orbit and beyond. Three years later, on 12th April, 1961, the Soviet Union put the first man into space, when cosmonaut Yuri Gagarin orbited the Earth – an accomplishment that was followed by the US sending Alan Shepard on a suborbital trip. Less than 50 years later, the two major space powers have been joined by a third, which has declared its intentions of not only establishing a space station, but also landing its astronauts on the Moon and eventually embarking upon a manned mission to Mars.

While the international media's attention to China's space program has been sporadic and sometimes patronizing, such indifference risks overlooking the long-term consequences of China as a growing space power and, more ominously, the possible confrontation of the US and Chinese interests in space. The recent successful manned spaceflights by Beijing and the bold predictions made by China have prompted some Western observers to wonder whether China's achievements signal the beginning of the end of the American dominance in manned spaceflight, while other analysts suggest that the rise of China's space program may represent the "Sputnik shock" all over again.

Perhaps more worrying than a race to the Moon are the potential political and militaristic implications of China's space ambitions. These aspirations are fuelled by aggrieved nationalism deeply ingrained in the Chinese psyche and a mindset dictating that China must develop economic wealth and military power so that it can exact retribution from the foreign powers that have humiliated China since the Opium War more than a century ago. Perhaps Beijing's pursuit of a robust and long-term space program is a rational decision to not only pay homage to this obsessive Chinese nationalism, but also to garner political and military benefits.

Against this background, the aim of this book is first to provide an overview of China's and the US's military and manned spaceflight capabilities. The second aim of the book is to consider the reality that the world faces a very different space race from the one pursued by the Soviet Union and the US in the late 1950s and 1960s. The final goal of the book is to consider the geostrategic implications of a new international rivalry that seeks to control the final frontier and how the capabilities of the adversaries may affect the outcome.

Acknowledgments

In writing this book, I have been fortunate to have had my wife, Doina Nugent, as my proof-reader. Once again, she has applied her considerable skills to make the text as smooth and coherent as possible. Any remaining shortcomings are my responsibility and mine alone.

I am also grateful to the five reviewers who made such positive comments concerning the content of this publication and to Clive Horwood and his team at Praxis for guiding this book through the publication process. The author also gratefully acknowledges John Mason, whose attention to detail and patience greatly facilitated the publication of this book. Thanks also to Jim Wilkie for creating the cover of this book and to the valuable care and attention of Christine Cressy and BookEns during the editing and typesetting process.

Once again, no acknowledgment would be complete without special mention of our cats, Jasper and MiniMach, who provided endless welcome distraction and entertainment.

About the author

Erik Seedhouse is an aerospace scientist with ambitions to become an astronaut. After completing his first degree in Sport Science at Northumbria University, the author joined the 2nd Battalion the Parachute Regiment, the world's most elite airborne regiment and greatest fighting force. During his time in the "Para's", Erik spent six months in Belize, where he was trained in the art of jungle warfare and conducted several border patrols along the Belize–Guatamala border. Later, he spent several months learning the intricacies of desert warfare on the Akamas Range in Cyprus. He made more than 30 jumps from a Hercules C130 aircraft, was certified in the art of helicopter abseiling, and fired more light anti-tank weapons than he cares to remember!

Upon returning to academia, the author embarked upon a Master's degree in Medical Science at Sheffield University. While studying for his Master's degree, he earned extra money by winning prize money in 100-km ultradistance running races. Shortly after placing third in the World 100-km Championships in 1992 and setting the North American 100-km record, the author turned to ultradistance triathlon, winning the World Endurance Triathlon Championships in 1995 and 1996. For good measure, he also won the inaugural World Double Ironman Championships in 1995 and the Decatriathlon, the world's longest triathlon – an event requiring competitors to swim 38 km, cycle 1,800 km, and run 422 km. Non-stop! Returning to academia once again in 1996, Erik pursued his Ph.D. at the German Space Agency's Institute for Space Medicine. While conducting his Ph.D studies, he still found time to win Ultraman Hawaii and the European Ultraman Championships as well as completing the Race Across America bike race. In 1997, *GQ Magazine* nominated the author as the "Fittest Man in the World".

Deciding it was time to get a real job, Erik retired from being a professional triathlete in August, 1999, and started work on his post-doctoral studies at Vancouver's Simon Fraser University's School of Kinesiology. While living in Vancouver, Erik gained his pilot's license, started climbing mountains, and took up sky-diving to relax in his spare time. In 2005, Erik worked as an astronaut-training consultant for Bigelow Aerospace in Las Vegas and wrote "Tourists in Space", the training manual for spaceflight participants. He is a Fellow of the British

Interplanetary Society and a member of the Aerospace Medical Association. Recently, he was one of the final 30 candidates of the Canadian Space Agency's Astronaut Recruitment Campaign. Erik currently works as a manned spaceflight consultant and author. He plans to travel into space with one of the private spaceflight companies. As well as being a triathlete, skydiver, pilot, and author, Erik is an avid scuba diver. Erik spends as much time as possible in Kona on the Big Island of Hawaii and at his real home in Sandefjord, Norway. Erik lives with his wife and two cats on the Niagara Escarpment in Canada.

Figures

Tables

Panels

Abbreviations

AAAS	American Association for the Advancement of Science
ABM	Anti-Ballistic Missile
ACS	Attitude Control System
AFRL	Air Force Research Laboratory
AFSPC	Air Force Space Command
AI	Artificial Intelligence
APSCO	Asia–Pacific Space Cooperation Organization
ASAT	Anti-satellite
ATB	Astronaut Training Base
ATCO	Ambient Temperature Catalytic-Oxidation
ATSP	Apollo Soyuz Test Project
ATV	Automated Transfer Vehicle
BACC	Beijing Aerospace Command and Control Center
BMD	Ballistic Missile Defense
BMDS	Ballistic Missile Defense System
BPC	Boost Protective Cover
C&C	Command and Control
C3	Command, Control, and Communications
C3PO	Commercial Crew and Cargo Program
CAIB	Columbia Accident Investigation Board
CALT	China Academy of Launch Vehicle Technology
CaLV	Cargo Launch Vehicle
CAST	China Academy of Space Technology
CAT	COTS Advisory Team
CAV	Common Aero Vehicle
CCB	Common Core Booster
CCDH	Command, Control, and Data Handling
CCP	Chinese Communist Party
CCS	Counter Communications System
CDMA	Code Division Multiple Access

CDV	Cargo Delivery Vehicle
CENTCOM	Central Command
CEV	Crew Exploration Vehicle
CIA	Central Intelligence Agency
CLV	Crew Launch Vehicle
CM	Crew Module
CMC	Central Military Commission
CMRS	Carbon Dioxide and Moisture Removal System
CNP	Comprehensive National Power
CNSA	China National Space Administration
CONUS	Continental United States
COSPAR	Committee on Space Research
COSTIND	Commission on Science, Technology, and Industry for National Defense
COTS	Commercial Orbital Transportation Services
CSA	Canadian Space Agency
CSSS	Constellation Space Suit System
CTU	Central Terminal Unit
CVO	Cargo Variant of Orion
DARPA	Defense Advanced Research Project Agency
DDTC	Directorate of Defense Trade Controls
DFH	Dongfanghong
DMSP	Defense Meteorological Satellite Program
DNI	Directorate of National Intelligence
DoD	Department of Defense
DRM	Design Reference Mission
DSCS	Defense Satellite Communications System
DSP	Defense Support Program
EAFB	Edwards Air Force Base
EAGLE	Evolutionary Aerospace Global Laser Engagement
ECLSS	Environmental Control Life Support System
EDS	Earth Departure Stage
EELV	Evolved Expendable Launch Vehicle
EKV	Exoatmospheric Kill Vehicle
ELINT	Electronic Intelligence
ELV	Expendable Launch Vehicle
EPIRB	Emergency Position Indicating Radio Beacon
EPS	Electrical Power System
ERS	Earth Remote Sensing
ESA	European Space Agency
ESAS	Exploration Systems Architecture Study
ESMD	Exploration Systems Mission Directorate
ET	External Tank
EVA	Extravehicular Activity

FALCON	Force Application and Launch from the Continental United States
FoS	Factor of Safety
FTV	Flight Test Vehicle
GBI	Ground-Based Interceptor
GOX	Gaseous Oxygen
GPS	Global Positioning System
GR&A	Ground Rules and Assumptions
GTO	Geostationary Transfer Orbit
HCV	Hypersonic Cruise Vehicle
HEMP	High Altitude Electromagnetic Pulse
HEO	Highly Elliptical Orbit
HPB	Horizontal Processing Building
HPUC	Hydraulic Power Unit Controller
HST	Hubble Space Telescope
IAF	International Astronautical Federation
IASS	International Association for the Advancement of Space Safety
IHPRPT	Integrated High Payoff Rocket Propulsion Technology
IMU	Inertial Measurement Unit
INS	Inertial Navigation System
IRD	Interface Requirements Document
ISC2	Integrated Space Command and Control
ISR	Intelligence, Surveillance, and Reconnaissance
ISRD	ISS Service Requirements Document
ISS	International Space Station
ITAR	International Trade in Arms Regulations
IUA	Instrument Unit Avionics
IVA	Intravehicular Activity
JAXA	Japanese Aerospace Exploration Agency
JPL	Jet Propulsion Laboratory
JSC	Johnson Space Center
JSLC	Jiuquan Satellite Launch Center
KEASAT	Kinetic Energy Anti-Satellite
KEW	Kinetic Energy Weapon
KKV	Kinetic Kill Vehicle
KSC	Kennedy Space Center
LADAR	Laser Detection and Ranging
LAS	Launch Abort System
LCH4	Liquid Methane
LEO	Low Earth Orbit
LES	Launch Escape System
LLO	Low Lunar Orbit
LM	Long March
LSAM	Lunar Surface Access Module

MCS	Mission Control Station
MCTR	Missile Technology Control Regime
MIT	Massachusetts Institute of Technology
MKV	Multiple Kill Vehicle
MNF	Multinational Force
MOL	Manned Orbiting Laboratory
MPSS	Main Parachute Support System
NDC	National Defense Complex
NDIO	National Defense Industry Office
NMCC	National Military Command Center
NMD	National Missile Defense
NOAA	National Oceanic and Atmospheric Administration
NPR	Nuclear Posture Review
NPR	NASA Procedural Requirements
NRC	Nuclear Regulatory Commission
NSB	National Science Board
NSIRA	National Security Intelligence Reform Act
OMS	Orbital Maneuvering System
OST	Outer Space Treaty
OTV	Orbital Test Vehicle
PAEC	Pakistan Atomic Energy Commission
PAROS	Prevention of an Arms Race in Outer Space
PBAN	Polybutadiene Acrylonitrite
PCU	Power Control Unit
PLA	People's Liberation Army
PLAAF	Peoples Liberation Army Air Force
PMAD	Power Management and Distribution
PNT	Positioning, Navigation, and Timing
PV	Photovoltaic
R&D	Research and Development
RCS	Reaction Control System
RFS	Radio Frequency Spectrum
RMS	Remote Manipulator System
ROE	Rules of Engagement
RSB	Reusable Solid Rocket Booster
RSM	Reactive Satellite Maneouvre
RSS	Rotating Service Structure
S&T	Science and Technology
SA	Spacecraft Adapter
SAA	Space Acts Agreement
SAFER	Simplified Aid For EVA Rescue
SAGES	Shuttle and Apollo Generation Expert Services
SBIRS	Space-Based Infrared System
SBL	Space-Based Laser
SCA	Spacecraft Adapter System

SDI	Strategic Defense Initiative
SEI	Space Exploration Initiative
SIGINT	Signals Intelligence
SLF	Shuttle Landing Facility
SLV	Small Launch Vehicle
SM	Service Module
SM-3	Standard Missile-3
SMP	Strategic Master Plan
SMSC	Space and Missile Systems Center
SPAS	Shuttle Pallet Satellite
SRB	Solid Rocket Booster
SRBM	Short-Range Ballistic Missile
SRM	Solid Rocket Motor
SROE	Standing Rules of Engagement
SSA	Space Situational Awareness
SSC	Stennis Space Center
SSME	Space Shuttle Main Engine
SSN	Space Surveillance Network
SSO	Sun Synchronous Orbit
STEC	Science, Technology and Equipment Commission
STSS	Space Tracking and Surveillance System
TLI	Trans-Lunar Insertion
TPS	Thermal Protection System
TSLC	Taiyuan Satellite Launch Center
TT&C	Telemetry, Tracking and Control
UN COPUOS	United Nations Committee on the Peaceful Uses of Outer Space
UNIDIR	United Nations Institute for Disarmament Research
USAF	United States Air Force
USSTRATCOM	US Strategic Command
UV	Ultraviolet
VAB	Vehicle Assembly Building
VDC	Volt Direct Current
VPB	Vertical Processing Building
VSE	Vision for Space Exploration
WSLC	Wenchang Satellite Launch Center
XSCC	Xi'an Satellite Control Center
XSLC	Xichang Satellite Launch Center
XSS	Experimental Spacecraft System

Section I

High Frontier Politics

Following its third manned spaceflight in 2008, China now stands at the pinnacle of the international space hierarchy, alongside Russia and the US. But, while the flights of Yang Liwei and his fellow taikonauts have received much media attention, what is less well known is the military dimension of its space program. Here, in Chapter 1, the broader historical and political contexts within which the Chinese civil and military space programs have developed are explored and in Chapter 2, Chinese space policy is examined against the policies of the US, China's rival in the new space race.

1

Rising dragon

THE *WHY* AND *HOW* OF CHINA'S LONG MARCH INTO SPACE

In October 2003, a Long March 2-F (LM-2F) rocket launched Shenzhou 5 (Figure 1.1) and China's first taikonaut, Yang Liwei, into low Earth orbit (LEO). Although the flight lasted only one day, and decades separated China's first manned mission from those of the Russians and Americans, the event was significant, as it heralded China as only the third nation ever to develop an independent manned spaceflight capability.

Despite being a poor developing country with a per capita income of only $1,293, China has indicated its intention to launch a space station, to land its taikonauts on the Moon, and eventually to embark upon a manned mission to Mars – ambitions it characterizes as a "long march" into space. In common with China's historic Long March, in which Mao Zedong's retreating forces created an epic propaganda coup, Beijing intends to ensure that the "long march" into space will, in addition to setting the tone for China's future, be seen as another grand project, on a par with the Great Wall. But how did China, whose space ambitions had often been denigrated by the Western media before Liwei's historic flight, accomplish a technological feat previously achieved by only two other nations, and what are the forces driving the red dragon's ascent into space? The answers to these questions are presented in this chapter, which first examines the impetus behind China's nascent space program, before describing the events leading to China's arrival on the threshold of attaining the status of a space power.

THE WHY

Manned spaceflight is open to all nations willing to pay the financial and technological price of admission (Panel 1.1). An activity that is perhaps the most difficult and most prestigious of all human endeavors – manned spaceflight – confers

Figure 1.1 Launch of Shenzhou-5: China's first manned space mission. Courtesy CNSA.

many benefits on those states bold enough to accept the risks and heavy costs involved. The awesome technological obstacles that must be overcome to send humans into space demand that a state acquire revolutionary space technologies, procured either through independent national development programs or through assistance from others. More often than not, it has been the latter path that has been pursued by those hoping to join the spacefaring elite, which is why China's lone-wolf approach appears to be so impressive, especially when the economic challenges of the country remain significant.

Panel 1.1. Hazards of manned spaceflight

Manned spaceflight represents an extremely difficult technological feat, fraught with danger. For example, hazards during lift-off include malfunction of the boosters, thrust oscillation with the potential to tear the launch vehicle apart, improper stage separation and dynamic overpressures capable of disintegrating the spacecraft. Crew can be killed by rapid or explosive decompression, either during the ascent to orbit or during on-orbit operations. Orbiting spacecraft can be penetrated by micrometeorites, navigation errors can result in ballistic re-entries imposing severe *g*-loading on deconditioned crewmembers, and heat shields can fail during the extraordinary heat of re-entry, as evidenced by the *Columbia* tragedy.

China's motivation

Given the dangers described in Panel 1.1, the question is why *any* nation would risk investing billions of dollars in a manned space infrastructure. Another question is why China has pursued a manned spaceflight program when other nations, clearly technologically and economically stronger, have failed. The European Hermes spacecraft, for example, was cancelled by the European Space Agency (ESA) due to excessive cost, forcing ESA to conduct their manned spaceflight activities by hitching rides with the Americans onboard their Space Shuttle, or with the Russians onboard their Soyuz.

Although China's justifications for pursuing manned spaceflight are myriad, some are more significant than others. Beijing did not send its taikonauts into space because Chinese leader, Jiang Zemin, is a space visionary, eager to explore the Moon and Mars as part of some quest to extend the envelope of manned spaceflight. For Jiang, who attained his lofty position by successfully navigating the labyrinthine maze of Beijing's power structure, the decision to support China's manned space program was simply a calculated risk. Jiang understood that while space successes may be spectacular, failures had the potential to scuttle national goals and damage his credibility and prestige as leader. However, it was a risk worth taking because, in Jiang's mind, China, by pursuing a manned spaceflight program, could boost domestic pride, gain international prestige, increase economic development, and reap all the benefits the US acquired through the Apollo and Space Shuttle programs.

Redressing the balance

It has been suggested that one reason a nation would pursue a manned spaceflight program is that only those states fearful for their survival follow such a route. This

Figure 1.2 Apollo 8 crew leaving crew quarters on December 21st, 1968. Courtesy NASA.

fear, the argument posits, is generated by concerns regarding military confrontation by other stronger states. Such a state of affairs undoubtedly drove the American–Soviet space race in the late 1950s and early 1960s (Figure 1.2).

In the first space race, the Americans and Soviets each felt threatened by the other due to perceived military weakness, while China, possessing no nuclear weapons, felt even more intimidated. The security dilemma posed by the threat of the two nuclear superpowers forced China to acquire a nuclear arsenal to redress the military imbalance. China's pursuit of ballistic missiles eventually laid the foundation of its space program, as was the case with the Americans and the Soviets, although none of the nuclear states originally aimed to reach Earth orbit. However, once the Americans and Soviets had developed their nuclear arsenals, a means was required to demonstrate their respective missile capabilities, and so the space race was born. Turn the clock forward four decades and a similar situation can be seen with Beijing using its space program to demonstrate Chinese military technology. In the 1960s, the Americans and Soviets, launching manned rockets into orbit as a surrogate demonstration of their military capabilities, taught Beijing a valuable lesson on the military relevance of manned spaceflight. That this lesson has been well learnt is evidenced by the sustained efforts of the Chinese to develop the necessary technological base required to accomplish military and commercial tasks presently conducted in LEO.

International prestige

In techno-nationalist terms, none of the regional manned spaceflight contenders such as India and Japan has even come close to achieving the technical feat of launching humans into orbit. Apart from placing China alongside the spaceflight elite, the accomplishment of launching taikonauts into LEO confers a significant leadership connotation, as evidenced by all the congratulatory telegrams from around the world following Yang Liwei's historic flight, perhaps most notably that of Indian Prime Minister, Atal Bihari Vajpayee. Although many countries offering their congratulations later downplayed the significance of China's first manned spaceflight, Beijing had at least established the perception of being Asia's space technology leader, and had regained what it considers as its rightful place among the technology leaders of the world.

Nationalism

China's manned spaceflight program strongly supports the country's national aspirations and is regularly used as a public instrument for securing China's stature on the world stage. Following the flights of Shenzhou 6 and Shenzhou 7, for example, China's taikonauts were transformed into political vehicles to promote an orgy of nationalist propaganda to publicize Beijing's rising influence as an independent and self-sufficient nation. The propaganda presented the missions as *made in China* from start to finish – a claim that ignored the fact Chinese taikonauts had received initial training in Russia in the 1990s. Similarly, the claim that China is independent and self-sufficient is nothing short of ludicrous, since the Shenzhou, in common with China's economic and technological development, is largely derivative. The brutal reality is that China functions as a huge cheap-labor platform for the global economy, and is completely dependent on foreign capital and its associated imported technologies, such as Russian space hardware. The reality of the *made in China* achievement promoted so vigorously by Beijing means subsistence wages, atrocious working conditions, unrelenting political repression, inefficient use of energy, and Armageddon-scale pollution.

Furthermore, while national prestige is certainly one of the driving forces behind China's manned spaceflight program, the claims of the Beijing leadership that the program is proof of an independent China is nothing but an illusion. Despite the carefully scripted comments of China's taikonauts, describing how manned spaceflight is rejuvenating the nation, the reality of endemic political repression, subversion, and corruption suggests these claims are nothing but a myth.

Economy

Many observers have raised the question of why Beijing wants to send taikonauts into orbit when China has problems feeding and clothing its people. Given the economic challenges of rampant unemployment, extreme rural poverty, insolvent state banks, uncontrolled corruption, and China's penchant for environmental annihilation, it would surely seem more sensible to give priority to economic development and improve people's quality of life.

Through the eyes of the leadership in Beijing, however, the rewards provided by pursuing a manned space program may provide the economic wherewithal to address some of the aforementioned problems. According to the International Space Business Council's *State of the Space Industry*, a report published in 2005, revenue from the space industry's global commercial services and government contracts is expected to surpass $158 billion by 2010. The ratio of financial input to output of the space industry is approximately 1:2, so a manned space program contributes to the goal of economic development lying at the center of Beijing's national development strategy.

In addition to the revenue generated by the sale of satellite launches, the deployment of weather and communications satellites provides platforms for tasks such as surveying crops, locating mining deposits, and measuring water resources. Already, satellite TV transmissions cover 8% of the population, and distance learning education has benefitted 20 million people. China's communications satellites have significantly improved disaster weather forecasting, reducing losses by several billion yuan every year, while its satellite communication network will have an annual value of 20 billion yuan, with more than 100 million people in remote areas benefitting.

The manned space program also has the potential to affect China's domestic economic environment. A good example of how funding for the space program has led to spin-off products entering the Chinese domestic market is the Outer Space Cup, manufactured by the Shanghai Wensu Industry Trade Company, Ltd. The cup was designed to withstand high temperatures and to be leak-proof, thereby alleviating obvious problems in microgravity. Following Yang Liwei's return, the Outer Space Cup sold millions, demonstrating how even a low-tech item from the

Figure 1.3 China plans to mine the Moon of Helium-3. Courtesy NASA.

space program can strengthen the economy. The penetration of spin-off technologies such as the Outer Space Cup into the domestic market is likely to continue in the future. For example, China eventually plans to mine the Moon of Helium-3 (Figure 1.3) – an economic endeavor with the potential to make large profits for the space program.[1]

Scientific and technological advancement

The manned space program has undisputedly greatly enhanced China's own technology and capability in space, but not all the great strides made in science and technology have been the product of hard work and self-reliance. While China is slowly closing the technology gap between itself and the original spacefaring nations, it has done so not only by pursuing bilateral and multilateral ventures, but also through more disreputable means.

Although China has made great efforts to develop a solid science and technology foundation (S&T) based on indigenous technologies, Russian-bought technologies and nefariously acquired American technologies also play a significant role in China's space program (Panel 1.2).

Political progression

Political influences in the space program play an important role in Beijing's desire for international prestige and recognition. The Chinese consider the attainment of an

Panel 1.2. China's Cold War spying

In February, 2008, a former Boeing engineer was arrested in the US for allegedly stealing Space Shuttle secrets for China. Dongfan Chung, one of four people taken into custody in a case of alleged espionage for China, was a naturalized US citizen employed by Boeing and was charged with stealing trade secrets relating to the Space Shuttle and the Delta IV rocket for China's benefit.

international standard of sophistication, such as placing a man in orbit, as a step towards becoming a world contender in the manned space arena – an achievement that will lead to recognition of China as a space power, or even a space superpower.

International relations

The speed and direction of China's manned space program are largely dictated by Beijing's ability to maintain international relations with countries willing to sell space hardware. To date, foreign acquisition from and international relations with Russia have played a decisive role in the pace of progress of the Chinese manned space program. Since the 1960s, as long as Russia has been willing to sell its space hardware and expertise, Beijing has been willing to buy it – an arrangement that shows no sign of changing in the near future. This is hardly surprising, since, to the Russians, maintaining a good relationship with Beijing means a steady flow of money into their space program and the potential for joint ventures with the Chinese.

Pakistan is another key partner that China relies upon to further its space ambitions. While the Sino–Russian relationship is legitimate, Beijing's relationship with Islamabad has caused concerns about US security. For example, when the US revealed illegal CSS-7 missile sales from China to Pakistan (Panel 1.3), the Bush Administration imposed sanctions on China. This event not only hurt China's space goals, but may have resulted in Beijing directing its space program towards a more militaristic route by viewing the US as a potential enemy.[2]

Social and cultural

As indicated in the State Council's 2000 *White Paper on China's Space Activities*,[3] space education is an important aspect of the space program, seeking to train space scientists and engineers, and foster space science interests among the student population. The success of the Shenzhou missions has already generated a significant increase in space interest among university students entering science and engineering

Panel 1.3. China's export of short-range ballistic missiles

In the early 1990s, Pakistan's National Defense Complex (NDC), a subsidiary of the Pakistan Atomic Energy Commission (PAEC), acquired complete though unassembled M-11s and possibly an undisclosed number of M-9 short-range ballistic missiles (SRBMs) from Beijing. Later, in the mid-1990s, according to the US Central Intelligence Agency (CIA), China apparently transferred an entire production line for M-11s and possibly M-9s to the NDC. Although China eventually agreed to abide by Missile Technology Control Regime (MTCR)* guidelines under US pressure, Beijing has not abided by the MCTR's key technological annex.

* The MCTR is a voluntary association of countries that share the goals of non-proliferation of unmanned delivery systems capable of delivering weapons of mass destruction, and that seek to coordinate national export licensing efforts aimed at preventing their proliferation.

departments, in addition to creating university programs focusing on more space-specific subjects.

The strategic high ground

Space activities are normally considered dual-use in nature, meaning the same space technologies that can lift a human into orbit can easily be used to deliver a warhead onto a target. As with the Americans and the Soviets in the late 1950s and early 1960s, Beijing's most important justification and motivation for pursuing a manned space program is based firmly in the military arena, which is not surprising, since national security remains a potent justification for the large expenditures demanded by a space program.

To that end, US space-based military assets have been routinely studied by the Chinese during the two Gulf Wars, and the campaigns in Kosovo, Afghanistan, and Iraq. From observing US military operations, such as *Desert Storm*, the Chinese soon realized that the military strength of the US was largely due to its advanced command, control, intelligence, surveillance, and reconnaissance abilities. These capabilities mostly rely on military satellites – assets the Chinese hope to match before employing their use in an attack on Taiwan (Panel 1.4). To achieve this goal, China is constructing a space-based surveillance infrastructure, including 20 differential global-positioning system stations to enhance the accuracy of the PLA's short-range ballistic missiles targeting Taiwan.

Panel 1.4. China's Taiwan policy

In the view of the PLA, the military power of the US, the potential to use that power to coerce China, and the ability to threaten China's pursuit of its own interests present a threat to China. Additionally, China's own threats against democratic Taiwan and the fact that PLA leaders believe the US is likely to come to Taiwan's assistance in the event of Chinese aggression against Taiwan magnify the perceived threat from the US.

The real *why* of China's spaceflight program

While international relations, political progression, and the other incentives cited in this section undoubtedly contribute to China's overall influence and provide Beijing with opportunities for international leadership, the true purpose of China's spaceflight program lies in the dual-use nature of space technology. Although Beijing is loathe to mention the military utility of its spaceflight program, the development of space hardware, combined with China's space doctrine, has several negative-sum aspects for the US, which may lead to future confrontation in space.[4] While many readers may be familiar with the recent successes of Beijing's manned spaceflight program, China's human space program and lunar exploration missions are intended to counteract concerns and divert attention from China's military uses of space. In reality, by striving to be a major space power, China has increased its comprehensive national power (CNP),* but its improving military space capabilities have resulted in the US viewing China as potentially coming into conflict with its own interests. The rise of China as a potential peer competitor raises concerns for the US, which, as we shall discover later, will increasingly define the rising dragon by military considerations, given the inherently military nature of the Chinese spaceflight program.

THE HOW

The Mao Zedong era

Even before the dawn of the space age in the 1950s, Mao and other Communist leaders were interested in joining the enterprise. At the time, China's motivation for

* CNP is defined as the sum total of the powers or strengths of a country in economy, military capabilities, science and technology (S&T), education and resources and its influence.[5] CNP may also refer to the combination of all the powers possessed by a country for the survival and development of a sovereign state, including material and ideational ethos, and international influence.[6]

pursuing a space program was strongly military-oriented, mainly as a result of the US threatening China with a nuclear attack if a truce in Korea was not established. China, without any credible deterrent to the American nuclear bomber forces, quickly searched for a strategic nuclear deterrent, leading Mao Zedong to pursue a space technology program. However, such a program took a long while to be realized, since, in the 1950s, China was extremely backward and its nascent technological infrastructure had been devastated by external and civil wars lasting almost 100 years.

Tsien Hsue-shen: the father of China's space program

Led by Tsien Hsue-shen (Panel 1.5), who had been to Germany as part of the American program to acquire Nazi rocket technology, China's early space program faltered at every step, dogged by one failure after another. In 1956, China managed to acquire two Soviet R-1 missiles – simple copies of the German V-2 rockets and woefully inadequate for developing an independent launch capability. A satellite project, started in 1958, was suspended in 1959 due to insufficient resources – a failure followed in 1960 by the successful flight of the DF-1 missile, a weapon derived from the old Soviet R-1. Meanwhile, millions of people in rural areas were dying of hunger but, due to the nuclear arms race between the US and the Soviet Union, China's sense of isolation was becoming ever more desperate, forcing it to pursue the development of missiles, whatever the cost. In 1960, China's pursuit of missile programs and satellite construction suffered another blow when, due to a personality conflict with Zedong, Nikita Khrushchev cut off China's only immediate source of technological support. Strong support from the state eventually overcame these constraints and the government allocated more resources to develop the DF-2, which was successfully tested in 1964. Once China acquired nuclear capability in October 1964, the DF-2 was redesigned to carry a nuclear weapon and was tested as DF-2A in October 1966 in the Xianjiang desert. The success of the DF-2A was quickly followed by the first DF-3 test, and the missile was redesigned so that Moscow could be brought within its range.

The route to China's first manned space program

In 1968, China's first manned space program was initiated under the National Defense Science Committee, which established the Space Medical Institute of China, tasked with research on manned spaceflight and the training of astronauts. During the program, 19 astronauts were selected from the Air Force, and the first launch of Shuguang-1 was planned by the end of 1973. Since China had yet to launch a satellite at the time, it wasn't surprising that the manned space program died shortly after.

Panel 1.5. Tsien Hsue-shen: father of the Chinese space program

The man who laid the foundation of the Chinese space program was a gifted scientist who worked for the US military in the 1940s and helped found the Jet Propulsion Laboratory (JPL).

Tsien Hsue-shen, also known as Qian Xuesen, created the Chinese space program from almost nothing at a time when his colleagues knew very little about rocket propulsion. Born in 1911, Tsien travelled to the US on a scholarship to study aeronautical engineering at the Massachusetts Institute of Technology (MIT) when he was 23. Shortly after arriving at MIT, he moved to Caltech to follow a path that led him to become one of the leading rocket scientists in the US. Under the tutelage of the Hungarian-American engineer and physicist, Theodore von Karman, Tsien became involved in rocketry and quickly became a star pupil at Caltech's Guggenheim Aeronautical Laboratory.

The 1943 discovery of German rocket activity resulted in the creation of JPL, with Tsien serving as research director, overseeing the development of solid-propellant missiles. By 1945, Tsien was working at the Pentagon with a high-level security clearance and writing reports on classified information and its implications for future military development.[7]

Following World War II, Tsien served as a member of the US technical mission that interrogated Nazi rocket scientists, among them Wernher von Braun, who was to become the father of the US space program.

An undisputed genius in the area of high-speed aerodynamics and jet propulsion, Tsien is credited with inspiring the 1950s Dyna-Soar project that ultimately led to the design of the Space Shuttle. At about the same time as Tsien was describing his idea for a spaceplane, Senator Joseph McCarthy began his campaign against widespread Communist infiltration, resulting in authorities revoking Tsien's security clearance. Although the Immigration and Naturalization Service had no evidence to support the charge, Tsien was a Communist, China was no longer a US ally, so, eventually, the immigration sought to deport him. Insulted by the actions of the immigration service, Tsien attempted to stay but, even with the support of Dan Kimball, Undersecretary of the Navy, circumstances conspired against the brilliant scientist, and he was finally deported in 1955.

From a historical perspective, Tsien served China well by administering its fledgling space program but, in terms of providing leading-edge technology, his impact is probably exaggerated because no single scientist, no matter how brilliant, can have more than a fraction of the knowledge necessary to design and develop launch vehicles. In fact, much of what he achieved as an administrator ultimately benefitted the US because China became an adversary within five years of Tsien returning to China, and missiles developed by the scientific complex he created were sent to the west of the country to bring Moscow within range. However, given that China is now a strategic rival of the US, Tsien's accomplishments are now more relevant.

Dongfanghong-1

Meanwhile, domestic political strife continued to impact China's space program, and it wasn't until 1970 that the country finally managed to launch its first satellite, *Dongfanghong-1* (DFH-1: Dongfanghong means "East is Red"), using a domestic launcher, the *Changzheng-1* (CZ-1: Changzheng means "Long March"), becoming the fifth country to achieve independent launch capability. However, it was another five years before China was able to launch a recoverable satellite, four years later than planned. The challenges of recovering a satellite from Earth orbit were significantly greater than simply launching a satellite, due to the need to overcome engineering difficulties such as a heat shield, a retrorocket system, and a control and ground tracking system.

Deng Xiaoping's era

A year after launching its first recoverable satellite, the Cultural Revolution came to an end with the death of Chairman Mao in September, 1976. A year later, in August, 1977, China's Four Modernizations program was accepted as the nation's new mantra. In sharp contrast to Mao's collectivism, which had only created poverty and despair, the Four Modernizations not only promoted the development of science and technology, but also reopened China to the outside world and started rebuilding relations with nations such as the US. Deng Xiaoping's policy of opening the country to the outside world reenergized the space programs, although the manned space program, which resumed under the interim Hao Government, was ultimately suspended in 1980. Instead, China focused on developing communications, meteorological and science satellites, launching the *Shijian-2* science satellite in 1981.

Developing the Long March launch vehicle

In parallel with its satellite development, China began development of the Long March LM-3 and 4 launch vehicles (Figure 1.4) and, in response to the US Strategic Defense Initiative (SDI) announced by President Reagan in March, 1983, the pace of the space program picked up significantly. The increased momentum resulted in the launch of a series of meteorological and science satellites between 1987 and 1994, their missions featuring various microgravity experiments, heavier payloads, longer stays on orbit, and increasing success, until the failure of a launch vehicle carrying the Australian *Optus-B2* communication satellite in December, 1992.

Setbacks

Reliability failures continued to plague China's nascent satellite business when, between 1995 and 1996, two separate commercial launches failed, causing wary

Figure 1.4 China's Long March launch vehicle. Courtesy CNSA (*see colour section*).

international players to consider Chinese rockets as being failure-prone. China's start-up satellite business was now confronted with the unforgiving realities of the marketplace and was forced to investigate the launch incidents. The investigative process involved US satellite manufacturers, marking a new degree of openness in Sino–American relations. The exchange of technical information concerning the technologies involved in the incidents required the Chinese to reveal their records, while the American companies were obliged to provide information pertaining to their payloads. Inevitably, the situation caused congressional concerns that proprietary information with dual-use capabilities was being released to China and an investigation was authorized by the US House of Representatives. In January, 1998, the outcome of the investigation revealed the accident investigation process had resulted in the Chinese acquiring technology that would enhance their military space capabilities, and severe restrictions were placed by the US Congress on any technology transfer that might threaten the US. The decision by Congress halted China's ability to compete for future US payloads and constituted a major setback.

Project 921

Shortly before the loss of *Optus-B2*, at the 1992 meeting of the International Astronautical Federation (IAF), the Chinese announced plans for a crewed capsule, indicating their renewed interest in manned spaceflight. Later that year, the State

Science and Technology Commission, the highest political level for S&T, revealed plans for a manned spacecraft by 2000, and a space station to follow shortly after. Finally, on September 21st, 1992, the roadmap for China's future in manned spaceflight was formalized as Project 921, the goals of which included a first crewed launch by 2002, an orbiting space laboratory by 2007, establishing a permanent space station, and launch of a first test spacecraft by 1998.

Project 921 hardware acquisition

Project 921 was designed around a new family of launch vehicles that could be built in a modular fashion. To meet the demands of the communication satellite market, China had been developing their modular LM-3A and 3B launch vehicles, which became operational in 1994 and 1996, respectively, but it was the development of the LM-2E and 2F launch vehicles that held the key to achieving the manned spaceflight goals of Project 921.

Debut of the Shenzhou capsule

The LM-2F offered a large lifting capability of nearly 10 tonnes into LEO and was designed to launch manned spacecraft, such as the Shenzhou capsule. Thanks to Russia's financial woes in the early 1990s, the Chinese were able to accelerate their manned spaceflight agenda by purchasing a Soyuz capsule emptied of flight instrumentation, a space suit, the Russian Kurs rendezvous system, and a docking module. Using this space hardware as a template, the Chinese developed the Shenzhou capsule (Figure 1.5), which was basically a recycled version of the Soyuz, albeit more robust. In parallel with the development of launch vehicles and spacecraft, the Chinese constructed four tracking ships and built new facilities at the Jiuquan Satellite Launch Center (Figure 1.6), including a vehicle assembly building, a transporter for moving launch vehicles, and a servicing tower on the launch pad.

Astronaut training

While China constructed launch facilities and acquired space hardware, the selection and training of taikonauts were underway. In common with the early selection campaigns of the US and Russia, Chinese taikonaut candidates were selected from experienced military fighter pilots with hundreds of hours' flight time. The selection campaign lasted two years and resulted in 12 candidates (Table 1.1) being selected from nearly 1,000 candidates.

Figure 1.5 China's Shenzhou manned space capsule. Courtesy Wikimedia.

Figure 1.6 Mission Control of Jiuquan Satellite Launch Center. Courtesy CNSA.

Table 1.1. Chinese taikonaut candidates.

Name	*Selection*	*Notes*
Quan Chen	January, 1998	Commander, Shenzhou 7
Qingming Deng	January, 1998	Pilot (active)
Jùnlóng Fèi	January, 1998	Commander, Shenzhou 6
Haipeng Jing	January, 1998	First back-up Flight Engineer, Shenzhou 6
Qinglong Li	November, 1996	One of two taikonauts sent to Russia for cosmonaut training
Bó Ming Lìu	January, 1998	First back-up Commander, Shenzhou 6
Wang Liu	January, 1998	Pilot (active)
Hǎishèng Niè	January, 1998	Flight Engineer, Shenzhou 6; second back-up Shenzhou 5
Zhanchun Pan	January, 1998	Pilot (active)
Jie Wu	November, 1996	One of two taikonauts sent to Russia for cosmonaut training; second back-up Flight Engineer, Shenzhou 6
Liwei Yang	January, 1998	First taikonaut in space; Commander Shenzhou 5
Zhìgāng Zhái	January, 1998	First back-up Shenzhou 5; second back-up Commander Shenzhou 6

Unmanned missions

The financial commitment to China's manned space program from its inception in 1992 to the first manned flight in 2003 is estimated at $2.1 billion – an amount that, by US standards, is small but, by China's standards, is huge. With the increased publicity associated with manned spaceflight and with international prestige on the line, China proceeded cautiously with the development and testing of its new spacecraft. However, despite the prudent approach, problems were encountered, most notably with the launch escape system (LES).* The temperamental Soviet system proved a handful for the Chinese, who embarked upon a series of tests to prove the reliability of the system. Additional problems were encountered in the development of the capsule, a space-rated version of which was scheduled for launch in 1999. However, while the LM-2F launch vehicle was ready for launch, the test version of the capsule was not. This forced the Chinese to upgrade the flyable prototype to flight status and fly it without instrumentation beyond basic guidance and recovery equipment. The launch of what was essentially a shell meant the Chinese were unable to evaluate flight effects upon test mannequins or undertake any experiments. The no-frills Shenzhou 1 flight went ahead on November 1st, 1999,

* The LES is simply a rocket motor that fires in the event of a launch anomaly, lifting the crew away from the launch pad and landing the capsule by means of parachutes.

flying 14 orbits, during which no on-orbit maneuvers were performed. Despite the unsophisticated mission architecture, the flight was plagued by problems such as failure to transmit commands to the onboard computer for initiating re-entry. Finally, on the final scheduled orbit, the problem was resolved, and the capsule landed on November 21st, 1999.

The next unmanned flight was launched on January 19th, 2001, and featured a more functional capsule including an operational orbital module. During the seven-day mission, Shenzhou 2 flew 108 orbits, and performed orbital maneuvers but, in common with the Shenzhou 1 mission, Shenzhou 2 was marred by problems with re-entry. Since no photos were released showing the returned vehicle, it was assumed some damage had occurred as a result of the parachute landing system having not deployed completely.

Having experienced two re-entry problems on successive flights, the Chinese knew the potential for humiliation and its adverse effect upon national prestige was very real. The re-entry phase of a mission is an extraordinarily complex and dangerous task, fraught with risks that may kill or injure the crew. The Soviets had experienced their share of hard landings in which cosmonauts had been injured and accidents in which crews had died. More recently, in February, 2003, NASA had lost the Space Shuttle *Columbia* during re-entry – a tragedy in which all seven crewmembers died.

The third unmanned mission flew to orbit on March 25th, 2002, carrying a mannequin and science experiments that included material and life science studies in addition to scientific payloads designed to perform atmospheric observations and space environment monitoring.

Manned missions

Finally, on October 15th, 2003, more than 42 years after Yuri Gagarin became the first man in space, 38-year-old taikonaut, Yang Liwei (Figure 1.7), blasted off into space atop a LM-2F rocket from the Jiuquan Satellite Launch Centre (JSLC). Attending the launch was Chinese President, Hu Jintao, and other senior leaders, but, despite the historic significance of the event, no live television pictures were permitted to be broadcast, since the Chinese leaders considered the political risk of a launch failure too great. Quoted shortly after the flight, Jintao praised the launch as "the glory of our great motherland", and described the flight as an "historic step of the Chinese people in the advance of climbing over the peak of the world's science and technology".

Less than two years later, on October 12th, 2005, Shenzhou 6 blasted off from the JLSC, carrying taikonauts Fèi Jùnlóng and Niè Hǎishèng for a five-day mission, reportedly costing US$110 million (by comparison, a typical Space Shuttle mission costs as much as US$500 million). Almost as significant as the orbital voyage of Jùnlóng and Hǎishèng was the fact the mission marked the 88th launch and 46th consecutive successful launches of China's LM series of launch vehicles.

Shenzhou 6 was followed, predictably, by Shenzhou 7, which shot up into an inky black sky on September 25th, 2008, carrying taikonauts Zhái Zhìgāng, Lìu Bó Mìng,

Figure 1.7 Yang Liwei, China's first taikonaut. Courtesy CNSA.

and Jing Haipéng. In another advancement in the space hall of fame for China, Zhìgāng performed the country's first extravehicular activity (EVA), spending 20 minutes outside the confines of the Shenzhou capsule. Following the taikonauts' return, Premier, Wen Jiabao, praised the mission, saying "The success of Shenzhou 7 manned space mission represents a historic step forward in our space technology. This is another success in China's manned space program. The country and its people will always remember what you have done".

SUMMARY

With its manned spaceflight capability, China has taken its place as a tier-one spacefaring nation, joining only the US and Russia in operating an independent manned space program. As a developing country, pursuing a full-spectrum space program that includes manned space missions, satellites, rocket design, and launch capabilities is a major feat. In the climate of great power competition, however, China's primary threat is the US, and America's exceptionalism stance is guaranteed to play a significant role in the new strategic battleground between East and West.

Having examined the background to China's space ambitions, the next step is to review the policy of the country's space program and compare this with the principles guiding the space program of the US. This will provide an insight into why the US and China recognize control of space as simultaneously a goal of and an enabler of military operations. It will also show how this will shape not only the future of warfare, but also the climate of space power competition in the manned spaceflight arena.

REFERENCES

1. "Manned Space Program Aims to Use Space Resources", *People's Daily* (October 14, 2003), internet version.
2. Sutter, P.A. China's Record of Proliferation Activities, *US–China Commission*. US Government Printing Office, Washington, DC (July 24, 2003).
3. "China's Space Activities", released by the Information Office of the State Council of the People's Republic of China, November 22, 2000.
4. US House of Representatives. *Report of the Select Committee on US National Security and Military/Commercial Concerns with the People's Republic of China*, Report 105-851. US Government Printing Office, Washington, DC (May 1999).
5. China Institute of Contemporary International Relations. *Global Strategic Pattern: International Environment of China in the New Century*. Shishi Press, Beijing, (2000).
6. Huang Suofeng. *New Theory on CNP: CNP of China*. China Social Sciences Press, Beijing (1999).
7. Chang, I. *Thread of the Silkworm*. Basic Books (November 1996).

2

US and Chinese space policy

THE CONDUCT OF NATIONAL SPACE ACTIVITIES IN THE PURSUIT OF HIGH AMBITION

At great expense, the US and China have developed the capability to not only launch people into space, but also to develop weapon systems capable of conducting warfare in space. The plan of action guiding the decisions concerning the civil and military space programs in the US are described in the 2006 US Space Policy document,[1] whereas China's civil space activities are described in a government White Paper, published in 2006.[2] This chapter provides an overview of these documents and discusses the space policies of each country.

US SPACE POLICY

On August 31st, 2006, President George Bush authorized a new US national space policy (Table 2.1) establishing principles governing the conduct of US space activities. The policy superseded Presidential Decision Directive/NSC-49/NTSC-8, National Space Policy of September 14th, 1996.[3] While previous policies had divided responsibilities between the Department of Defense (DoD) and the Director of Central Intelligence, the new policy demanded different considerations due to the creation of the Directorate of National Intelligence (DNI).

Principles of US space policy

Since its inception, the US space program has been guided by principles driven by the knowledge that those who effectively utilize space will enjoy added prosperity and security and will hold a significant advantage over those who do not. Given the potential of freedom of action in space to enhance national security and increase

Table 2.1. Sections of US national space policy.[3]

1	Principles	7	International Space Cooperation
2	Policy Goals	8	Space Nuclear Power
3	General Guidelines	9	Radio Frequency Spectrum & Orbit management & Interference Protection
4	National Security Space Guidelines	10	Orbital Debris
5	Civil Space Guidelines	11	Effective Export Policies
6	Commercial Space Guidelines	12	Space-related Security Classification

economic prosperity, the conduct of the US space program is a top priority, guided by fundamental principles.

In common with other spacefaring nations, the US has always stated its commitment to the exploration and use of space for peaceful purposes, and for the benefit of all humanity. However, inconsistent with the tenet of "peaceful purposes" is the pursuit of national interests by the US defense and intelligence agencies. This issue is addressed by the principle of *sovereignty* in which the US rejects any claims to control by any nation over space or celestial bodies. The principle also rejects any limitations on the fundamental right of the US to operate in and acquire data from space.

The policy's third principle addresses the issue of *cooperation*, which directs the US to cooperate with other nations in the peaceful use of space and to extend the benefits of space and space exploration. This principle is echoed by the Vision for Space Exploration (VSE), explicitly directing NASA to "pursue opportunities for international participation to support U.S. space exploration goals".[4] Perhaps the most visible embodiment of current US cooperation in space is the International Space Station (ISS) (Figure 2.1), a partnership between NASA, the European Space Agency (ESA), the Japanese Aerospace Exploration Agency (JAXA), the Canadian Space Agency (CSA), and the Russian Federal Space Agency.

As controversial as the principle of *sovereignty* is the policy's *right of passage* principle. This states that the US "considers space systems to have the rights of passage through and operations in space without interference".[1] This statement is comparable to the right to innocent passage across the world's oceans in international waters. Many left-leaning observers have expressed reservations concerning the real intent of this principle, worrying that it signals US intentions for space supremacy. In the context of security, however, the principle abides by the belief that just as the US needs a navy to enforce freedom of the seas, it also needs to be able to exert control over space.

The principle of *national interests* encompasses the US's space capabilities, including ground, space-based, and supporting elements. To uphold this principle, the US states that it will preserve its rights, capabilities, and freedom of action in space, while at the same time deter others from either impeding those rights or developing capabilities intended to do so. If necessary, the US also affirms that it upholds the right to protect its space capabilities and respond to interference and deny, if necessary, adversaries the use of space capabilities hostile to US national

Figure 2.1 The International Space Station is an example of nations cooperating in space. Courtesy NASA (*see colour section*).

interests. Additionally, in keeping with the policy of American exceptionalism, the US opposes the development of new legal regimes prohibiting or limiting US access to or use of space. Furthermore, the policy states clearly that proposed arms-control treaties must not impair the rights of the US to conduct, research, testing, operations, or other activities in space for US national interests.

Given the many benefits derived from the US space program, it is not surprising that the US is committed to encouraging and facilitating a growing commercial space sector, another key objective expressed in the policy. To that end, the US intends to use US commercial space capabilities to the greatest practical extent, consistent with national security.

The language of the principles governing US national space policy is unmistakably security-driven. Additionally, the policy stresses the belief that US control of space is not only essential to defend against attacks on the US and coordinate preventive attacks against adversaries, but also fundamental to US prosperity. Furthermore, the policy's principles have undoubtedly reinforced international perceptions that the US may choose to develop, test, or deploy space weapons. Such a situation clearly reflects an emphasis on the Air Force's freedom of action in space and the marking of an important step forward in a long-fought campaign by right-wing hawks to extend their agenda to space.

Table 2.2. US space policy objectives.[1]

- Strengthen the nation's space leadership and ensure space capabilities are available in time to further US national security, homeland security, and foreign policy objectives
- Enable unhindered US operations in space to defend US interests there
- Implement and sustain an innovative human and robotic exploration program with the objective of extending human presence across the solar system
- Increase benefits of civil exploration, scientific discovery, and environmental activities
- Enable a dynamic, globally competitive domestic commercial space sector in order to promote innovation, strengthen US leadership, and protect national, homeland, and economic security
- Enable a robust science and technology base supporting national security, homeland security, and civil space activities
- Encourage international cooperation with foreign nations and/or consortia on space activities that are of mutual benefit and that further the peaceful exploration and use of space

Policy goals and guidelines

To achieve its space policy objectives (Table 2.2), the US intends to embark upon a series of tasks, ranging from developing space professionals to increasing interagency partnerships. An overview of these objectives is presented here.

The role of science and engineering

Given that science has been one of the most important successes of NASA, it is not surprising that US space policy emphasizes a program focusing on science and technology (S&T). Scientific knowledge and the application of revolutionary technologies have been one of the most tangible products of America's investment in space. This achievement is reflected in a policy calling for encouraging an innovative commercial space sector and the provision of incentives for high-risk/high-payoff transformational space capabilities.

The US recognizes that sustained excellence in space-related science, engineering, and operational disciplines is key to the future of US space capabilities. However, ensuring a strong science and engineering workforce may prove a barrier to a bright future for the US space program, as evidenced by a recent observation by the National Science Board (NSB):

> "We have observed a troubling decline in the number of U.S. citizens who are training to become scientists and engineers."
>
> National Science Board, *Science and Engineering Indicators 2004*[5]

The projected shortfall in the US's science and engineering workforce and commensurate rise in science and engineering degrees among China's citizens (Figure 2.2) may result in the US facing challenges realizing other aspects of space policy such as space systems procurement. The drought in the science and engineering workforce harkens back to the dawn of the first space race, when the Russians beat the US into space with Sputnik in 1957, amid fears the US was falling behind in science and engineering. In response to these fears, the US determined to put a man on the Moon, and the space race began. The US invested in education, and ensured a generation of younger people would be inspired by the potential and possibilities emerging from the space race. But, in a dramatically short period of time, the US lapsed. Turn the clock forward 50 years and fears concerning lack of engineering and science expertise are being voiced again. In 1980, the US and China each graduated a similar number of engineers, but, by 2000, Chinese engineering graduates had increased 161% while US graduates had declined 20%. Since 2000, the trend has deteriorated further – a development that may seriously impede the competitiveness of the US in the race to the Moon.

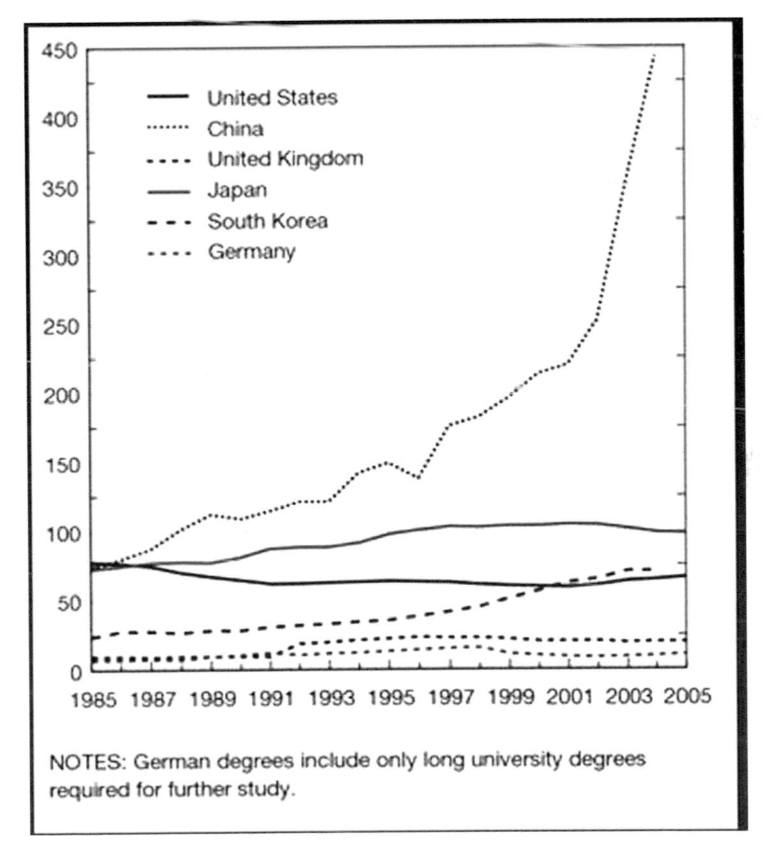

Figure 2.2 The comparative shortfall in engineering degrees in the United States. Courtesy National Science Board.

Figure 2.3 The Space Shuttle has been an integral enabling element in sustaining American manned spaceflight capability. Courtesy NASA.

The problems of ITAR

In addition to ensuring a strong science workforce, the US must also ensure space system development, procurement, *and* mission success. These goals are dependent on research, acquisition, management, execution, oversight, and operations. Collectively, they represent key factors enabling the space system development that has been fundamental to American success in space (Figure 2.3).

Despite its extraordinary record of space system development and its status as the unchallenged leader in delivering space capabilities, the US recognizes that there are challenges to its preeminence in the space industry arena, such as the increase in the number of satellites launched by China. To ensure its position as a leader in commercial space systems, the US is focusing on not only maintaining realistic and stable funding, but also strengthening interagency partnerships and opportunities for collaboration. Unfortunately, this goal is compromised by restrictive export control regulations.

Although US space policy states that the challenges of the 21st century demand a focused and dedicated unity of effort, US policy on export controls has seriously damaged its ability to negotiate agreements and joint ventures with foreign companies. A consequence of these restrictive regulations, which fall under the purview of the International Trade in Arms Regulations (ITAR, see Panel 2.1), is the present situation of the US satellite industry, which has great difficulty competing in the world market due to extraordinarily rigid export requirements.

Panel 2.1. Overview of the International Trade in Arms Regulations

The ITAR is administered by the State Department's Directorate of Defense Trade Controls (DDTC) and governs the export of defense articles from the US.

Those products qualifying as defense articles are listed in Part 121 of the ITAR, including items such as parts, accessories, attachments, and associated equipment specifically designed or modified to be used for military purposes. Also listed are any technical data directly related to any product designation with military applications. Making the ITAR even more restrictive is Category XXI, entitled "Miscellaneous Articles". Category XXI encompasses "any article not specifically enumerated in other categories which has substantial military applicability and which has been specifically designed, developed, configured, adapted or modified for military purposes".

Responsibility for designating products as defense articles is the task of the State Department, which follows the ITAR policy statement as a guideline for designating defense items. According to the ITAR, articles may be designated as defense items if they: (1) are specifically designed, developed, configured, adapted or modified for a military application, and do not have (a) a predominant civil application and (b) a performance equivalent to that of an article or service used for civil applications; or (2) are specifically designed, developed, configured, adapted, or modified for a military application and have significant military or intelligence applications.

Given ITAR's restrictions, even innocuous items such as a fairing on a launch vehicle will qualify as a defense item and will require the exporter to acquire an export license. The same applies to ITAR-controlled technical data, as long as the item/data is intended for a country not on the embargoed or prohibited list, such as Iran, North Korea, and, of course, China.

Why is ITAR so restrictive?

ITAR's complexity and restrictiveness are a result of the Reagan and the George H.W. Bush Administrations. In 1988, the Reagan Administration decided to permit the launch of American commercial satellites by China in exchange for establishing launch quotas and technology-safeguard agreements with Beijing. The commercial satellite launches became a valuable factor in obtaining non-proliferation agreements with Beijing and also served to liberalize the economic competitiveness of the US space industry. The launches were so beneficial that they continued throughout the George H.W. Bush Administration and into the Clinton Administration until two commercial satellites were lost in failed launches. The events and debacle that ensued precipitated an attack on the Clinton Administration's liberal export control policies. Worse was the dispute over US involvement during China's investigation of the cause of the launch failures, which led to accusations that US participation had somehow aided Beijing's ballistic missile program. The upshot of the fiasco was Congress passing legislation resulting in the sale of satellites as well as satellite technology becoming controlled as munitions and new restrictions were placed on the transfer of technology to China.

National security space guidelines*

Policy guiding US national defense space capabilities is set forth in the National Security Intelligence Reform Act (NSIRA) of 2004. The NSIRA provides responsibilities for the Secretary of State, in consultation with the Secretary of Defense and the Director of National Intelligence (DNI). The NSIRA also directs the Secretary of Defense to support the President and Vice President in supporting and enabling defense requirements during times of peace and crisis in addition to developing and deploying space capabilities that sustain US advantage. Furthermore, the NSIRA provides guidelines for employing strategies resulting in optimized space capabilities that support national security.

Achieving national defense policy objectives

To realize the national defense space policy, the Secretary of Defense is directed to maintain a multitude of capabilities. These capabilities range from the execution of space support, force enhancement, space control, and force application missions to providing capabilities to support continuous, global strategic and tactical warning. Included in the Secretary of Defense's purview is the development of capabilities to ensure freedom of action in space, and, if necessary, deny such freedom of action to adversaries.

Working with the Secretary of Defense to achieve the goals of national defense space policy is the DNI. The DNI, among other duties, provides intelligence collection and analysis of space-related capabilities to support space situational awareness for the US government and US commercial space services. The DNI also establishes and implements policies concerning the operational details of intelligence activities related to space.

---

* While the 2006 US space policy document establishes national security guidelines for the use of space, it does not establish fundamental principles by which the US military is guided in their actions in support of national objectives. This particular aspect of military space affairs is found in US doctrine, which forms the basis for Chapter 3.

Figure 2.4 The 2004 Vision for Space Exploration aims to return astronauts to the Moon by 2020. Courtesy NASA.

Civil space policy

In January, 2004, President George W. Bush announced the VSE – a bold venture that included returning humans to the Moon by 2020 (Figure 2.4), and embarking upon a manned mission to Mars by 2035. However, with the near demise of the US satellite industry due to export controls, the squeezing of NASA's budget over several years, and an increasingly go-it-alone approach, there appear to be serious obstacles to realizing the VSE's lofty goals. Given these current difficulties, combined with a recession predicted to extend deep into 2010, US space policy appears to be a contradiction of elevated ambition and fading commitment.

The US civilian space program was, and is, characterized by scientific exploration and discovery, using human and robotic means. NASA's robotic missions to various destinations in the solar system have produced a revolution in scientific understanding of the planets, asteroids, and Earth's immediate environment. Similarly, the manned spaceflight program has had an equally remarkable history of accomplishment, ranging from Apollo to Shuttle–Mir and from the ISS to Constellation.

US civil space policy is coordinated by the Secretary of Commerce, through the Administrator of the National Oceanic and Atmospheric Administration (NOAA) in coordination with the NASA Administrator. Together, these officials are responsible for continuing to consolidate US preeminence by planning and executing challenging space science and exploration missions. For example, civil space policy tasks the NOAA and NASA to continue the program of civil geostationary operational environment satellites. The policy also assigns these agencies to conduct programs of

Figure 2.5 SpaceX's *Dragon* capsule may help NASA reduce the five-year hiatus in manned spaceflight capability following the end of Shuttle operations in 2010. Courtesy SpaceX (*see colour section*).

research to advance scientific knowledge of the Earth through space-based observation and development of enabling technologies.

Commercial space guidelines

The myriad achievements of US space science and exploration are also inextricably linked to the success of the commercial space industry. Since US companies find themselves at a serious competitive disadvantage internationally, due to the ITAR restrictions, US policy tends to focus more on using national commercial space capabilities rather than pursuing foreign commercial services. An example of this policy is the Commercial Orbital Transportation Services (COTS) program (Panel 2.2), initiated by NASA on January 18th, 2006, to coordinate the commercial delivery of cargo and crew to the ISS. The outcome of the COTS program was NASA entering into contracts with Orbital Sciences and SpaceX (Figure 2.5) at the end of 2008 to utilize commercial cargo vehicles.

Panel 2.2. NASA Commercial Crew & Cargo Program

A major tenet of US space policy is the pursuit of commercial opportunities for providing transportation and other services to support the ISS and eventual exploration missions to the Moon. To that end, the NASA Commercial Crew & Cargo Program Office (C3PO) at the Johnson Space Center (JSC) was established. The C3PO's vision is to extend human presence in space by enabling a robust US commercial space transportation industry. To realize this objective, the C3PO aims to implement US space exploration policy with investments to stimulate the commercial space industry. It also aims to facilitate US private industry demonstration of cargo/crew space transportation capabilities with the goal of achieving reliable, cost-effective access to low Earth orbit (LEO).

The C3PO's strategy creates privately owned and operated space transportation systems, with NASA serving as a lead investor and customer of transportation services. To implement this strategy, NASA has established a two-phased approach, Phase 1 being the COTS Demonstrations project. Under COTS, NASA helps industry to develop and demonstrate its own crew and cargo space transportation capabilities, while industry leads and directs its own efforts, with NASA providing technical and financial assistance. A good example of this is the relationship between NASA, SpaceX, and the Orbital Sciences Corporation (Figure 2.6). Phase 2 is the ISS Commercial Resupply Services, in which the government is conducting a competitive procurement for cargo services to support the ISS.

Under the agreement with NASA, the COTS commercial partners are responsible for the overall design, development, manufacturing, testing, and operation of their COTS system. To be eligible for NASA financial assistance, the COTS partners must also conduct the COTS demonstrations according to the terms and conditions set out in a document known as the Space Act Agreements (SAA). They must also abide by the COTS ISS Service Requirements Document (ISRD), the COTS Human Rating Plan, and the ISS to COTS Interface Requirements Document (IRD). NASA, for its part, monitors the progress of its commercial partners through an assessment of the SAA milestones and provides expert technical assistance as requested or where considered necessary via the NASA COTS Advisory Team (CAT).

Given the impending competition with China, the COTS program may prove a vital component not only in reducing the interruption in manned spaceflight capability, but also in enabling the US to return to the Moon within the timeframe promised by the VSE.

Figure 2.6 Orbital Sciences Corporation's Taurus launch vehicle lifts off from Vandenberg Air Force Base. Courtesy OSC.

International space cooperation

One of the most important questions plaguing the VSE is the extent to which other nations will be invited to join the US as it embarks upon missions to the Moon and Mars. While President Bush appeared to invite other nations to participate in the VSE when he announced the program, the global perception in the aerospace industry is that the US has little interest in bringing other nations into the planning process. Such a path appears contradictory, since US policy clearly states intent to pursue international cooperation on space activities that are of mutual benefit and that further the peaceful exploration and use of space. Such a path is also inconsistent with the route followed historically, since the Space Shuttle and ISS programs were only accomplished with considerable international involvement and exchange of data.

Perhaps the issue most threatening to US policy on international cooperation is the growing international perception that the US intends to control space militarily. Such a stance is sure to impact the progress that has been made over the last five decades towards multilateral international cooperation.

Space nuclear power

The US develops and uses space nuclear power systems in circumstances that enable or significantly enhance space exploration and/or operational capabilities. For example, the Cassini vehicle (Figure 2.7), launched in 1997, utilized nuclear propulsion – a decision requiring the US to implement certain safety measures. These safeguards are implemented by the US Government, the Secretary of Transportation, and the Nuclear Regulatory Commission (NRC), which collectively license activities prior to launch and conduct safety analysis and nuclear safety monitoring.

Radio frequency spectrum and orbit management and interference protection

The issue of radio frequency spectrums (RFSs) and orbital assignments has become increasingly important, as the amount of orbital traffic has escalated since the publication of the 1996 space policy document. To ensure continued access to RFSs (Panel 2.3), the US aims to assure that commercial, civil, and military space activities are not subject to interference that may compromise space capabilities. To that end, the policy seeks spectrum regulatory status under US domestic regulations for those satellites owned and operated by the government.

Figure 2.7 Cassini spacecraft. Courtesy NASA.

Panel 2.3. Radio frequency spectrum

International and national management of the RFS has grown to be an exceedingly complex function, since geopolitical, legal and economic factors, as well as technical factors, influence the development of radio regulations. Consequently, RFS management is inextricably bound to governmental policy and regulation.

The technical factors fundamentally stem from mutual interference considerations that impose certain constraints on users of the RFS. Over the past decade especially, the rapid increase in spectrum use by space telecommunications services has significantly increased the mutual interference problems extant among space services and between space and terrestrial services. Since the geostationary orbit (GTO) is highly desirable for many space services, the utilization of this resource has become an international concern.

Orbital debris

Given the potential for space debris (Figure 2.8) to cause havoc in LEO, it is not surprising that US space policy seeks to preserve a safe space environment. Under the auspices of the US Government Orbital Debris Mitigation Standard Practices,

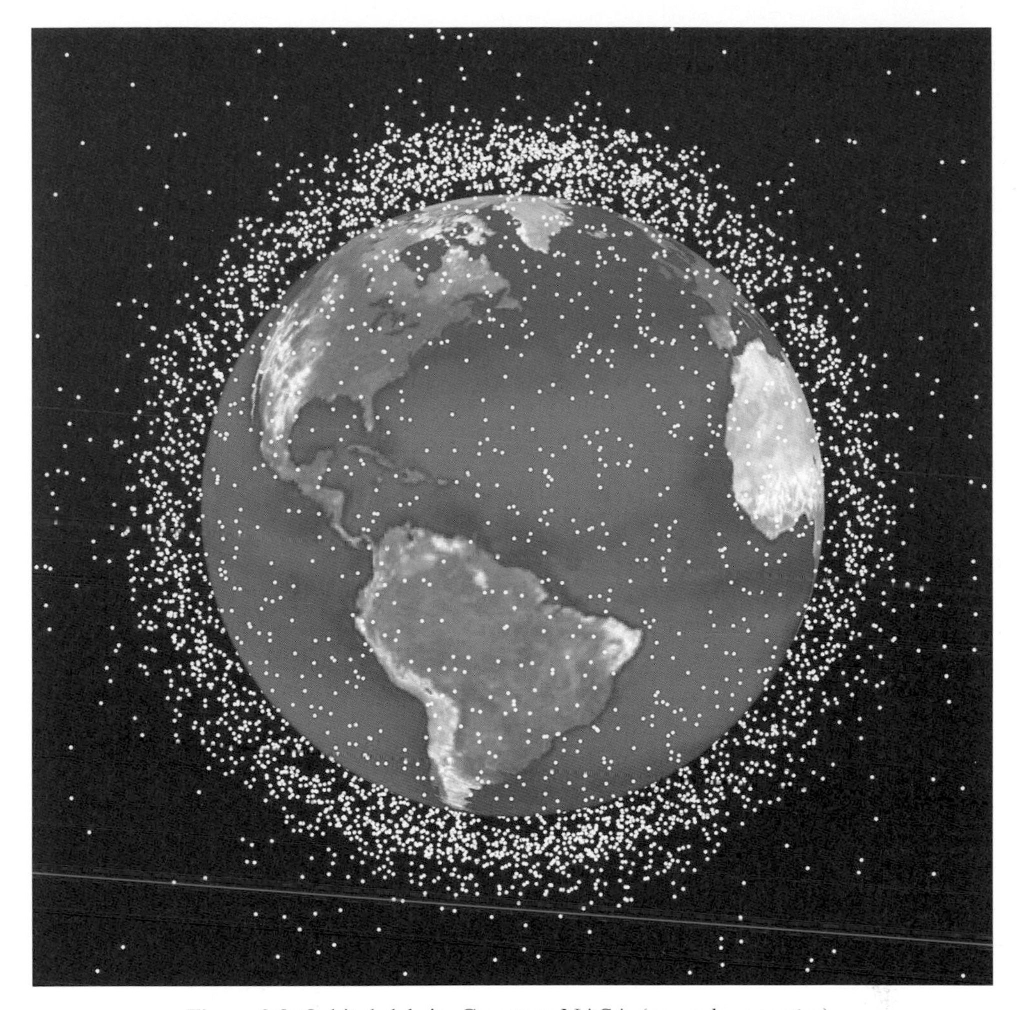

Figure 2.8 Orbital debris. Courtesy NASA (*see colour section*).

policies and practices are implemented seeking to minimize debris by assessing mission requirements and the operation of spacecraft.

Analysis of US space policy

Since its publication, many left-leaning journalists have attacked the new policy, claiming it suggests the US is seeking space superiority and intends to deny access to space to countries that the US deems unacceptable. However, while some may interpret this to be the case, the policy, while using forceful language, does not state that the US intends to be the world's space law enforcement agency. Rather, US policy merely states it will not accept a situation "whereby other countries can deny

America access to space".[1] In reality, there is a big difference between acting as the world's space police and indicating you will not permit other countries to push you around! More temperate analysts and journalists have suggested that, in terms of actual policy positions, the new policy is not much different from the one it replaced.

The new policy begins by declaring that the "The United States is committed to the exploration and use of outer space by all nations for peaceful purposes, and for the benefit of all humanity".[1] It goes on to say that the "The United States will seek to cooperate with other nations in the peaceful use of outer space to extend the benefits of space, enhance space exploration and to protect and promote freedom around the world".[1] These two principles have been long-standing US policy goals for more than five decades, and underline the US commitment to the peaceful use of space. However, due to the weighty tone of the document, it is easy to ignore the policy's peaceful intentions, and focus instead on the unilateral approach to access to space. For example, analysts have noted that whereas the 1996 policy listed five objectives for the US space program and mentioned national security for two of them, the 2006 policy lists six objectives and cites national security in four of them. Equally troubling to some analysts and critics is that the policy neither specifically endorses the deployment of space weapons nor does it make clear that the administration is opposed to such actions. The increased importance of security and defense in space is further underscored by the unilateralist tone of the new policy, which makes it clear that the administration is wary of an arms threat to America in space and will consider appropriate policy positions on arms control to enhance the security of the US. The increased significance of space defense is also highlighted in references to the word "cooperation" in relation to international space activities. Whereas the 1996 policy document mentioned *cooperation* a dozen times, the new policy alludes to it only four times. A similar comparison can be made with the phrase "arms control", mentioned seven times in the 1996 document, but only twice in the 2006 policy. This process of counting the number of times a word or phrase is used is a standard technique for determining the importance that certain policies have in official documents. A content analysis of the 2006 space policy clearly points to a de-emphasis of international cooperation and arms control. This de-emphasis not only reflects a fundamental change in ideology and political control, but also indicates the change in issues confronting the US space program compared with more than a decade ago. While the civil space sector received clear direction from the White House in 2004 in the form of the VSE, the issue of space security is in need of greater direction, a circumstance reflected in the new policy. A good example of this is the inclusion in the 2006 document of a section on access to RFS and orbit management and interference protection, which was not included in the 1996 policy.

On the subject of policies concerning the civil space program, the 2006 document does not stray far from the goals of the VSE. However, it indicates clearly that exploration is not NASA's sole goal, stating that the NASA Administrator shall "execute a sustained and affordable human and robotic program of space exploration and develop, acquire, and use civil space systems to advance fundamental scientific knowledge of our Earth system, solar system, and universe".[1] Surprisingly, the 2006 policy makes no mention of the ISS or the Space Shuttle, which were both mentioned in the VSE document. Another change compared with the 1996 policy is the attention devoted to Earth science and observation – a subject to which an entire section is devoted in the new document compared with a scant six mentions in the 1996 version. Another noteworthy aspect of the 2006 policy is the section on space nuclear power, which is significantly longer than the 1996 version, despite the cancellation of the civilian Prometheus* program:

> "When I was growing up, NASA united Americans to a common purpose and inspired the world with accomplishments we are still proud of. Today, NASA is an organization that impacts many facets of American life. I believe NASA needs an inspirational vision for the 21st Century. My vision will build on the great goals set forth in recent years, to maintain a robust program of human space exploration and ensure the fulfillment of NASA's mission. Together, we can ensure that NASA again reflects all that is best about our country and continue our nation's preeminence in space."
>
> Barack Obama

While the 2006 space policy document serves an important function in guiding the activities of the US space program, certain aspects of it have already become moot since the Obama Administration took office and the onset of the global economic recession. However, President Obama made human space exploration a major part of his science platform and, at the time of writing, it appears his administration will support President Bush's VSE. On space security matters, President Obama's policy is less clear. For example, one option suggested by the incoming administration was to tear down the long-standing barriers between NASA and the Pentagon in an effort to accelerate the VSE amid the prospect of a new space race with China. Regardless of what changes the Obama Administration decide upon, space policy will continue to be a prominent and contentious public policy issue, particularly as it relates to national security, technology, and space exploration. While it may appear that US space policy is well defined, in reality, the connected facets of the US space program are integrated in the complex world policy arena in which serious challenges threaten the uncertain future of America's space program. Of all the challenges to US preeminence in space, perhaps none is greater than the threat posed by China, whose recent accomplishments in the military and manned spaceflight arena have resulted in profound diplomatic echoes.

CHINESE SPACE POLICY

As mentioned in Chapter 1, Beijing's space policy is largely driven by the need to add to China's comprehensive national power (CNP), defined as the sum of a nation's

---

* Project Prometheus was established by NASA in 2003 to develop nuclear-powered systems for interplanetary missions. In 2005, it was reduced to a low-level research effort, and eventually cancelled.

economic, political, educational, scientific, technological, and military strength. CNP can be separated into *hard power*, such as the armed forces, and *soft power*, such as societal and economic influence. While its space program is not considered a primary contributor to China's CNP, it certainly plays an important role given space activities increase its hard power by improving military capability.

Overview of China's space policy

China's space policy is manifested in its pursuit of space power and its ability to portray China as a nation committed to the peaceful uses of space while simultaneously serving Beijing's political and military interests. Engaging in a space program not only provides China with opportunities for international cooperation, but also increases its influence at regional and international levels. On the surface, pursuing such a policy would be considered a natural motivation for conducting an active space program. However, since the anti-satellite (ASAT) incident in January 2007 and China's abstention at the UN Security Council following North Korea's failed satellite launch in April 2009,* the real impetus of China's space pursuits is undeniably linked to military utility.

China's space program is a two-tiered system – one that Beijing would lead you to believe is akin to the US civilian space program, and the other that is intended for military uses of space. In reality, there hardly seems to be a reason for a two-tiered system, since the PLA remains in control of the *entire* Chinese space program. Given the ambiguity, lack of transparency, and elusive nature of the PLA, the intentions of Beijing's space policy are notoriously difficult to deduce. Since there is no government document outlining Chinese space policy, an insight into Beijing's future intentions in space can only be deduced by analyzing the forces influencing China's space program and deciphering intentions in Beijing's Five-Year Plans and occasional White Paper.

Forces influencing China's space policy

The dominant force influencing China's space policy is the PLA, which understands the importance of space control and space denial for future wars. To that end, the PLA is intent on ensuring the development of cutting-edge technologies that will enable China to enhance its war-fighting capabilities in space. According to the PLA, space is now considered one of China's strategic frontiers. This position drives Beijing's space policy in the pursuit of acquiring offensive counterspace systems and working on technology for the purposes of attacking foreign satellites.

---

* Following North Korea's failed satellite launch on April 5th, 2009, the United Nations Security Council stated that the launch violated a council resolution and called for sanctions against the reclusive nation. China opposed the resolution.

In addition to the PLA, international relations and foreign science and technology undoubtedly influence China's space policy. Clearly, the most powerful exogenous influence is the US, whose position of space superiority directly impacts the direction of China's space program, as evidenced by the PLA's militaristic efforts. Other influences include Chinese relations with Russia and the ESA. These joint ventures not only play a decisive role in the speed with which China's space program evolves, but also influence the amount of money flowing into the space program.

China's strategy

In recent years, China has made significant progress across a broad range of space technologies, including launch vehicles, satellites, and human spaceflight, in addition to taking a leading role in regional space cooperation. Although its technology still lags behind the major space powers such as the US, Europe, and Russia, China's burgeoning space program provides opportunities for Beijing to use the benefits derived from its space activities to become a more respected nation. Although China does not have a grand space strategy akin to the VSE, it does publish Five-Year Plans, from which it is possible to deduce Beijing's intentions for its space program. Generally, these Five-Year Plans portray China as a modernizing nation committed to the peaceful use of space while simultaneously serving Beijing's political, economic, and military interests. While the statements pertaining to the peaceful uses of space are little more than fiction, China's space program does confer important political, commercial, and military benefits. Its commercial activities in the field of space technologies and launch services, for example, increase revenues for the space industry as well as advancing China's diplomatic interests with oil-rich countries. Furthermore, the expansion of its international cooperation on space activities heralds China's increasingly influential foreign policy. Domestically, by developing a high-profile manned space program, the Communist Party demonstrates that it is the best organization to launch China to its rightful place in world affairs.

Five-Year Plans

Beijing's space program, as well as China's economy, is governed by a series of economic policy decisions generated every five years. These Five-Year Plans document specific goals spanning every sector of the Chinese economy, from coal-mining to construction, and from steel manufacturing to space exploration. Presently, Beijing is in its Eleventh Five-Year Plan, governing the period from 2006 to 2010:

> "Our country is one of the few major space powers. China's position in the world and the country's security depend on the continued fast development of space technology."
>
> COSTIND press release announcing its Eleventh Five-Year Plan

Table 2.3. Goals of China's Eleventh Five-Year Plan.

	Goals
Manned spaceflight	• Increasing mission complexity, including extravehicular activity, rendezvous and docking, and establishing a space station
Lunar exploration	• Launch lunar orbiter to take three-dimensional images of the Moon • Conduct surface exploration of the Moon
Space science	• Develop X-ray modulating telescope to study black-holes • Launch Shijian-10 recoverable satellite to conduct space biomedicine experiments • Develop solar telescope to study solar activity
International cooperation	• Participate in Sino–Russian Mars environment exploration plan • Participate in the Sino–French Small satellite Solar Flare Exploration Project
Imaging satellites	• Launch remote sensing satellites, including all-weather, multi-spectral, and differential-resolution Earth observation

According to the Commission on Science, Technology, and Industry for National Defense (COSTIND),* the Eleventh Five-Year Plan (2006–2010) includes five primary science projects (Table 2.3).

White Papers

In addition to the information published in the Five-Year Plans, some insight into China's space policy can be gleaned from White Papers, such as the one published by the Information Office of the State Council on October 12th, 2006. Entitled *China's Space Activities in 2006*, the document is divided into the following five sections: "Aims and Principles of Development", "Progress Made in the Past Five Years", "Development Targets and Major Tasks for the Next Five Years", "Development Policies and Measures", and "International Exchanges and Cooperation".

According to the 2006 White Paper, the broad aims of China's space activities (Table 2.4) are based on guiding the development of the country's scientific and technological programs, making innovations independently, and achieving leapfrogging development in key areas such as launch technology. The White Paper also

* COSTIND was established in 1982 by the merger of the Defense Science and Technology Commissions (DTSC), the National Defense Industry Office (NDIO), and the Science, Technology and Equipment Commission (STEC) of the Central Military Commission (CMC).

Table 2.4. Broad aims of China's space activities.

1.	Promote human civilization and social progress for the benefit of mankind
2.	Utilize space for peaceful purposes
3.	Enhance understanding of the Earth and the cosmos
4.	Meet the demands of economic construction, scientific and technological development, national security and social progress
5.	Raise the scientific quality of the Chinese people
6.	Protect China's national interests and rights, and build up the comprehensive national strength

emphasizes that China will protect the space environment and develop and utilize space resources in a rational manner. However, less than a year after the paper was published, China committed the ultimate act of environmental vandalism by conducting its ASAT test, turning low Earth orbit (LEO) into a veritable minefield of thousands of objects travelling at 27,000 km/h.

A key section of the White Paper is the third, which outlines two sets of objectives to be achieved by 2010 and 2020. The development targets include improving the capabilities and reliability of launch vehicles, completion of a satellite remote-sensing application system, and the establishment of a satellite navigation and positioning system. In the manned spaceflight arena, China aims to engage in extravehicular activities (EVAs), a feat it achieved in September 2008, and to achieve spacecraft

Figure 2.9 China aims to conduct research onboard an autonomously orbiting space laboratory. Courtesy CNSA.

rendezvous and docking, a skill required for a manned lunar mission. Another objective is to conduct research onboard a short-term manned, and long-term autonomously orbiting, space laboratory (Panel 2.4 and Figure 2.9) – a goal predicted to be realized by 2010.

Panel 2.4. China's space station

China is developing Tiangong, a 17,000-lb man-tended military space laboratory planned for launch by late 2010 – a mission that will coincide with phase-out of the Space Shuttle. The project is being led by the General Armaments Department of the People's Liberation Army, and will give China two separate station development programs. Shenzhou 8, the first mission to the station, will be flown unmanned to test robotic docking systems. Subsequent missions will be manned to utilize the new pressurized module capabilities of the station. In a development sure to cause concern among Western defense analysts, China is openly acknowledging that the new Tiangong station will involve military space operations.

The design includes a large module with a docking system making up the forward half of the vehicle and a service module section with solar arrays and propellant tanks making up the aft. The concept is similar to the European Space Agency's (ESA) manned concept for the Automated Transfer Vehicle (ATV). While Tiangong will be used as a target to build Chinese docking and habitation experience, the station's military mission has some parallels with the US Air Force Manned Orbiting Laboratory (MOL) program, cancelled in 1969.

MOL's objectives were primarily reconnaissance and technology development, but, while US military astronauts were to be launched in a Gemini spacecraft atop their MOLs, in China's case, the module will operate autonomously and be visited periodically by Chinese taikonauts. In addition to achieving space policy and operational mission objectives, the Chinese station will surely guarantee a propaganda windfall for Beijing and send a global geopolitical message relative to the decline of US space leadership.

The development policies, described in the fourth section of the White Paper, are divided into the three fields of space technology, space application, and space science. Not surprisingly, many of the measures in each of these fields support manned spaceflight, such as the construction of launching and operational services and the transformation and development of space technology.

Key policy elements

Manned spaceflight

Beijing's manned spaceflight program is by far its most high-profile space endeavor. However, until China successfully launched its taikonauts into orbit, international attention on its space program had been sporadic and patronizing at best. The media either denigrated China's space program or treated it nonchalantly, mainly because China's entry into the manned spaceflight arena had come so late. Having finally achieved the goal of manned spaceflight, the focus of China's space industry over the next few years will be to strengthen its innovative capabilities and find ways to develop the industry faster. Evidence of these objectives is statements announcing the establishing of a space station, and preparing for an eventual manned mission to the Moon.

China's statements of embarking upon manned lunar missions have made many US officials nervous, especially given the aggressive pace at which the Chinese space program is progressing. These anxieties are compounded by the knowledge that if China were to succeed in landing taikonauts on the Moon ahead of the return of American astronauts, not only would Beijing acquire enormous international prestige, but US leadership in space would be in doubt. Given these concerns, it is difficult not to link the US's renewed enthusiasm for space, as embodied in the VSE, to the current race against China's rapid rise in space.

Science and technology

The centerpiece of China's space science program is the Chang'e robotic lunar exploration program. Chang'e, launched in 2007, is a precursor mission to a soft landing mission planned tentatively for 2012 and a lunar sample return mission in 2017.

Another prominent project is the Double Star satellite program, the result of an agreement between the China National Space Administration (CNSA) and ESA on July 9th, 2001. Designed to research the effects of the Sun on the Earth's environment, Double Star comprises two Chinese satellites and four ESA satellites, which together form a monitoring network.

Although China's space program pursues various research and development programs, many of the accomplishments of the PLA's space policy have been achieved by pursuing a strategy of skipping generations of technology. This has been achieved either through robust bilateral joint ventures with Great Britain, France, and ESA, space-related acquisitions from Russia, or by nefarious means, as described in Chapter 1. Given the success of this strategy, it is likely the PLA will continue to fill technology gaps by these means, especially given China's lunar ambitions, which will require technologies the PLA have yet to acquire.

Satellite export

China's first satellite export agreement was signed in December, 2004, between the Nigerian Government and the China Great Wall Corporation. The agreement required China to build and launch the satellite, provide operating services, and train Nigerian technicians in its operation. Based on China's *Dongfanghong-4* communication satellite, the Nigerian Communication Satellite was launched on May 14th, 2007.

Cooperation

Despite conducting ASAT tests and opposing UN Security Council resolutions, China takes a positive role in activities organized by the United Nations Committee on the Peaceful Uses of Outer Space (UN COPUOS). China has also acceded to the "Treaty on Principles Governing the Activities of States in the Exploration and Use of Outer Space, Including the Moon and Other Celestial Bodies", and, ironically, the "Convention on International Liability for Damage Caused by Space Objects". China has also actively participated in activities organized by the Inter-Agency Space Debris Coordination Committee, and even played host to the 36th Committee on Space Research (COSPAR) Scientific Assembly in July, 2006.

China has long expressed interest in both regional and international cooperation in space. This interest is evidenced by China's endorsement of the Asia–Pacific Space Cooperation Organization (APSCO) convention in October, 2005, promoting multilateral cooperation in space science, technology, and application, and its cooperative efforts with ESA. China's cooperation with ESA is primarily in the field of Earth observation in which ESA promotes the use of its data from its Earth Remote Sensing (ERS) and Envisat satellites to develop Earth observation data of land, ocean, and atmospheric conditions.

China also agreed to invest in the European *Galileo* satellite navigation and positioning system. However, the *Galileo* agreement lost momentum when European businesses decided to develop much of the technology themselves, and concerns that US export controls might not have permitted the use of certain US technologies on *Galileo* due to the possibility of their diversion to China. With their cooperation with Europe reduced, China has increased its collaboration with Russia, as evidenced by 29 new projects being added to the cooperation program in 2005.

China's level of cooperation may be further reduced as a consequence of China's nascent military space program and its ASAT test, which gave rise to serious doubts globally about China's true intentions in space, especially those nations that already have space satellites! Most worrying to those countries cautious about cooperating with China was the fact that the Chinese military did not hold prior consultations on the test with other government agencies such as the foreign ministry and security establishment.

Economic effects

The annual budget for the entire Chinese space program, estimated to be between \$1.8 and \$2.2 billion, is less than one-twentieth of the combined NASA and Pentagon space budgets. However, given the opaque nature of China's information dissemination and the fact that China's COSTIND does not provide an annual assessment for the space budget, this figure is speculative. Despite such a small budget, the PLA is still able to pursue a full-spectrum space program that belies its funding, thanks in part to its S&T strategy. Given its comparatively small funding, the goals of China's space program are obviously not the same as the US. However, by pursuing an innovative space policy based on a narrow technological base, China has created a stand-alone space capability. Rather than develop a wide spectrum of space activities, as the US has done over the past 40 years, China's strategy focuses on specific disciplines that it hopes can match and perhaps out-perform the US. Many might argue that any new space race is a race for second place, but the burgeoning Chinese space program is already beginning to surpass the achievements of ESA, which has no independent manned spaceflight capability, but leads the world in launching commercial satellites.

China's space policy analyzed

The wish-list described in the White Paper is as impressive as it is ambitious, but questions remain as to whether China has sufficient resources to accomplish these tasks. Furthermore, the White Paper makes no reference to military applications of the space program, dwelling instead on the principles of exploration and utilization of space for peaceful purposes – statements that, in light of the ASAT test, are nothing more than hot air.

US and China's space strategies

While we may not know much about the character of Beijing's space policy, we can gauge China's progress in space. Furthermore, it can be asserted definitively, as evidenced by the language of the VSE and the principles cited in the 2006 policy document, that the US is determined to maintain by all means possible their preeminent position in space.

With a continuation of Beijing's political will demonstrated thus far in the civil space arena, many observers argue that China may soon begin to challenge the US in some space activities. In fact, with the retirement of the Space Shuttle in 2010, and a possible five-year hiatus in US manned spaceflight capability, some observers may posit that China may eventually lead the new space race. However, given the encouraging developments in the American commercial space industry, it is likely that the interruption in manned space capability will be significantly reduced. Furthermore, while it may appear China's space program is developing in leaps and bounds, the reality is their level of space experience is more than four decades behind the US's. However, while China's manned spaceflight ambitions may cause some anxiety among US space officials, it is the challenge to US military dominance in space that is perhaps the most ominous aspect of the new space race and it is this subject that is the topic of the next section.

REFERENCES

1. www.fas.org/irp/offdocs/nspd/space.pdf.
2. http://news.xinhuanet.com/english/2006-10/12/content_5193446.htm.
3. www.globalsecurity.org/space/library/policy/national/nstc-8.htm.
4. www.nasa.gov/pdf/54868main_bush_trans.pdf.
5. National Science Board. *Science and Engineering Indicators 2004*. National Science Foundation, Arlington, VA (2004), www.nsf.gov/sbe/srs/seind04/start.htm.

Section II

Dark Arena

SPACE WARFARE AND CHINA'S THREAT TO US SPACE SUPERIORITY

The US is the world's foremost space power today, but this position is not assured in perpetuity. Of all the nations in the world, the US is the most reliant on space, and is therefore the most vulnerable to the disruption of its space assets – a weakness China fully intends to exploit in the event of a conflict. Furthermore, the US's quest for full spectrum dominance in the space arena represents a power tactic challenging China's core national interests. Given the US threat to China's security, it is hardly surprising that Beijing's military doctrine is shaped to counter the US effort. A recent example of this doctrine was China's anti-satellite (ASAT) test in January, 2007, which represented something of a wake-up call for the US. Furthermore, China's reckless act in low Earth orbit (LEO) represented a high-leverage, asymmetric threat with the potential to inflict a highly disproportionate impact on US military capability and security. Since many US space-based assets serve both civilian and military users, their destruction, and even the threat of their destruction, could have devastating economic and military consequences, ultimately wreaking havoc on the US and global economy. Against this background, it is inevitable concerns are being raised by military theorists and space analysts. For example, "Is a space doctrine emerging in China, and if so, what are its contours?"; "Is China developing a preemptive strategy?"; and "What is the role of deception in Chinese military space strategy?" Chapter 3 addresses these questions while steering clear of the blogosphere-based misinformation that seems to seethe around the subjects of space doctrine and strategy.

While it is necessary to establish a doctrine for fighting in the harsh and unforgiving space environment, the best national strategy in the world is of no value without space assets, without which doctrine cannot be implemented. The advanced space hardware of the US comprises a complex network of space-based command, control, communications, and surveillance and reconnaissance capabilities that form the key to American combat operations, as evidenced in Operation *Desert Storm*.

These assets, however, are relatively soft and mostly defenseless, and, while they embody the very nature of American military might and power, they are also the source of deep vulnerability – a weakness the Chinese military recognizes. To that end, the Chinese are developing conventional weapon systems designed to disable American satellites and destroy US ground stations. In Chapter 4, US and Chinese space hardware is described and comparisons made between current and future space weapon systems, ranging from American and Chinese ASAT capabilities to direct attack and directed-energy weapons.

Given the inordinate American dependence on its space assets and the perceived asymmetric advantage of China's counterspace program, the US is pursuing a strategy aimed at responding to asymmetric warfare by continuing to utilize its military dominance to deter and defeat adversaries. This tenet of *space dominance* is addressed in Chapter 5, which explains how the US will defend the High Frontier and how China's intentions to match the US may ultimately and inevitably fall short.

3

Space warfare doctrine

> "Space assets provided exhaustive information about the enemy's status and measures he was taking. By thoroughly knowing the status of Iraqi troops, the MNF [Multinational Force] command paralyzed their operations and stunned them with the unexpectedness of steps being taken. In the future the role of space in war evidently will rise sharply, since the capabilities of strategic means of warfare are realized to the maximum extent in the aerospace sphere. It is presumed that in the not too distant future unavoidable strikes by precision weapons and weapons based on new physical principles can be delivered from space against any targets regardless of their degree of hardening. Thus, a country not having the capability to counter space weapons may turn out to be doomed."
>
> Major General I.N. Vorobyev[1]

The phrase "space weapons" is almost guaranteed to cause an unfavorable reaction in the US, whose population generally views space as the purview of NASA. The truth is that space weapons capability is the 21st-century way of war. Such a capability promises those nations who control space the capacity to use force to influence events around the world in a timely, effective, and sustainable manner. No event better illustrated this power than Operation *Desert Storm*, the world's first true space war, which took place in 1991.

Operation *Desert Storm* was a conflict characterized by unfathomably complex warfare equipment, missile interceptors, stealth aircraft, and other elements of space-supported and space-enabled forces. Thanks to its space-based assets, the US decimated the world's fourth largest military in just 10 days of ground combat. It was an achievement that would have been impossible without support from space. Twelve years later, *Iraqi Freedom*, the sequel to *Desert Storm*, underlined the central role of space power, when the US stepped over the threshold of a new way of war.

As warfare evolves to the point at which it is imperative to neutralize important targets quickly, the deployment of space weapons (Figure 3.1) to defeat space threats is inevitable and an arms race in space unavoidable. Although space continues to be viewed by many as a place to be exploited purely for peaceful purposes and the benefit of mankind without national claim or jurisdiction, in reality, space is already

Figure 3.1 Raytheon's Lightweight Exoatmospheric Projectile (LEAP) is a highly modular, lightweight, space-tested interceptor element for the Standard Missile-3. Courtesy Raytheon (*see colour section*).

militarized. Very soon, space will become just another war-fighting environment, and guiding the deployment and use of weapons in this environment is *warfare doctrine*. But what is doctrine? In this chapter, the concept of warfare doctrine is articulated in the context of how military affairs are, and will be, governed in the space environment.

FOUR SCHOOLS OF SPACE DOCTRINE

There is no common definition of "doctrine", even among military professionals who profess to understand it. Some definitions describe doctrine as a set of rules or principles governing the employment of military forces, but most military officials agree it can best be thought of as a set of beliefs:

> "Military doctrine is what is officially believed and taught about the best way to conduct military affairs."
>
> Professor I. B. Holley[2]

While Holley's definition is a useful starting point, in reality, the concept of doctrine is a little more complicated, since, when it applies to space, there are four identifiable belief structures, each of which has different assumptions of how to best employ space assets.

Sanctuary school

The precept of this school is that the primary value of space assets is their capability to conduct reconnaissance within the boundaries of sovereign states – a principle based on a spacecraft's over-flight capabilities. Advocates of the *sanctuary* school doctrine (Panel 3.1) contend that any arms limitations treaty cannot be achieved without space systems being employed as a technical means of treaty compliance.

Panel 3.1. Sanctuary doctrine: the deterrent strategy

For a state to be successful in implementing the *sanctuary* doctrine, it must first be able to absorb a first strike and still be capable of inflicting unacceptable damage upon the aggressor. This deterrent strategy is grounded in the belief neither side will permit the other to acquire sufficient weapons to make the first strike so damaging that the other's retaliatory forces do not survive. As witnessed during the Cold War, the downfall of the *sanctuary* doctrine is that it leads to arms buildups in the absence of arms limitations agreements. However, where the *sanctuary* doctrine really falls apart is the belief that the doctrine's core is dependent on space being considered a sanctuary, completely absent of military assets. The final nail in the coffin of the *sanctuary* doctrine came in January, 2007, with China's anti-satellite (ASAT) test – an event that arguably tainted the space environment forever.

The *sanctuary* doctrine was the official doctrine of the US during the Eisenhower Administration and subsequent administrations up to and including the Carter Administration. However, sanctuary beliefs were rendered practically null and void in September, 1982, with the establishment of Space Command[3]. Ultimately, the death knell for the *sanctuary* doctrine was President Reagan's announcement in March, 1983, for a renewed emphasis on ballistic missile defense (BMD)[4].

While the *sanctuary* doctrine was an appropriate strategy for the era of the Cold War, it was a strategy developed by peace-loving officials who saw the deterrent value of space. However, with the deployment and testing of ASATs and the development of space weapons ranging from exoatmospheric kinetic kill vehicles (KKVs) to space-based lasers (SBLs) (Figure 3.2), the *sanctuary* doctrine has long since gone the way of the dodo.

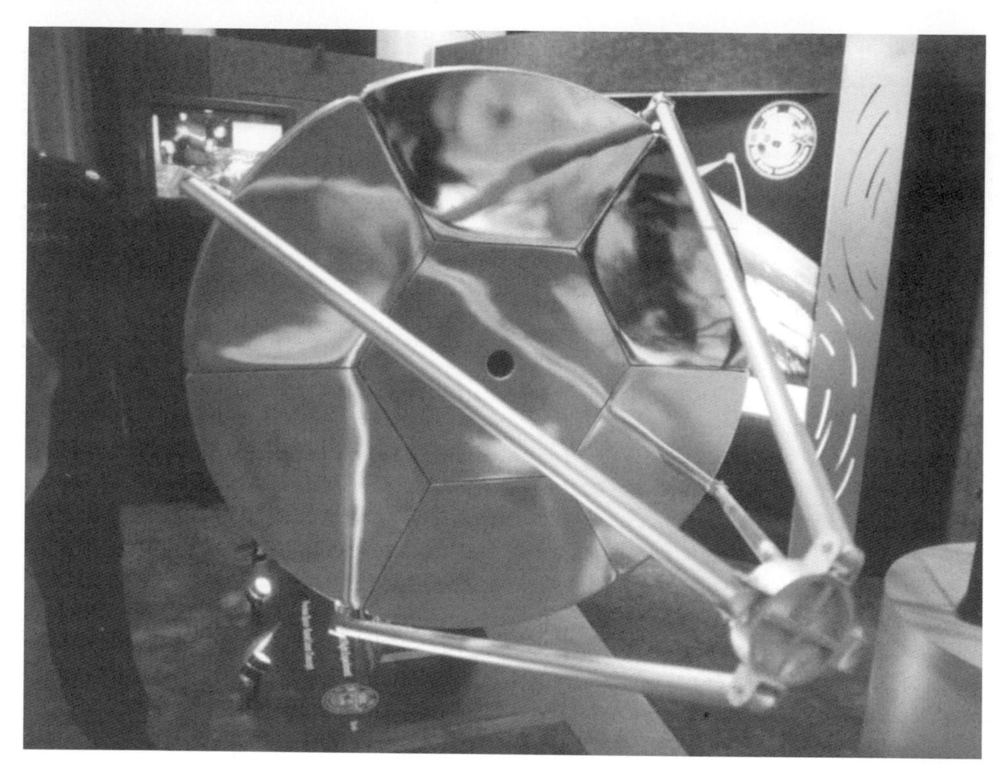

Figure 3.2 Lockheed Martin's space-based laser. Courtesy Lockheed Martin.

Survivability school

This school of thought is based on the *vulnerability* of space systems. The tenet of *survivability* proponents (Panel 3.2) is that space assets are less survivable than terrestrial forces. The doctrine is supported by the belief that nuclear weapons are more likely to be used in space than on Earth and the limited maneuverability of space assets results in decreased survivability. Because of the limited survivability of space assets, advocates of this doctrine argue that space forces should not be depended upon because they will simply not survive – a belief that has particular relevance to China's space warfare strategy.

The nuclear ASAT scenario has influenced both the US and China by fostering development of ASAT weapons designed to be used as part of a space denial strategy. This "retaliation-in-kind" strategy is appropriate when considering a nuclear exchange that would result in neither side surviving intact. While the *survivability* school fosters such a strategy, it is not the best strategy for winning wars in space, and therefore has little value as a means of protecting and preserving military assets.

Panel 3.2. Survivability

The demise of the *sanctuary* doctrine gave birth to the *survivability* doctrine, a strategy that has become ever more significant as a result of China's ASAT capability. While China's ASAT carried no warhead, the payload could just as easily have been a nuclear weapon – a scenario bringing a new dimension to the vulnerability of space assets. The damage even a small nuclear payload could inflict upon satellites in orbit underscores the fundamental tenet of the survivability school that argues space assets are more vulnerable than terrestrially based assets.

Domination

Advocates of the *domination* school (Panel 3.3) are the same people who suggested a space-based BMD system, arguing that the high ground ensuring domination is the best offensive and defensive strategy. *Domination* proponents contend the global-presence capabilities of space assets such as directed-energy and ASATs provide opportunities for revolutionary national defense strategies capable of reversing any stalemate caused by other nations' offense capacity.

Panel 3.3. Domination

The domination school is synonymous with the "Star Wars" BMD system – a strategy that protects the US by space-basing an active defense system, thereby relegating the deterrent doctrine to the dustbin. The domination doctrine is often described by its supporters as a defensive strategy, known as *mutually assured survival*, despite an absence of a firm foundation of the technological robustness of the system. However, given the US preeminence in space and weapon systems, the technology gap is closing to a point at which many of the systems required to realize an active defense system are achievable.

While assured survival may be a doctrine valued by many strategists, once again, it is not a means of winning a war. To achieve a more effective state of deterrence, it is necessary to adopt an offensive/defensive strategy based on controlling the environment.

Control

The *control* school (Panel 3.4) advocates assign no value to space assets. This school believes that whichever nation has the capability to control space will also possess the capability to control the surface of the Earth. This belief structure argues that control of the space arena is a capability that will deter war and, in future conflicts, space control capability will be on a par with the ability to control the air and sea environments.

> "Our charter is to rapidly obtain and maintain space superiority and the space, nuclear and conventional strike capabilities that produce desired warfighting effects. This requires a fundamental shift in our thinking. Instead of focusing on the force enhancement role of our space systems and the deterrence role of our nuclear and conventional forces, we must also pursue the ability to apply conventional combat in, from, and through space."
>
> Excerpt from US Air Force Space Command's Strategic Master Plan, 2006

Panel 3.4. Control

Many of the themes among the terrestrial doctrines share a commonality with the basis for a space control doctrine. For example, it is generally accepted that control is a capability rather than a condition. A good illustration of this is the way navies control friendly sea lanes during times of conflict but do not actively deny other navies their sea lanes in peacetime. A similar control concept is applied by air forces, but the navy and air force way of control cannot be applied to space assets, since it is impossible to exclude other nations from the space environment, as space is infinite. Since space assets cannot anchor and occupy positions, space control must be exerted over certain strategic areas of space such as geostationary orbits.

ENFORCING A SPACE POWER DOCTRINE

From an assessment of the four doctrines, it is evident that the most effective way to employ space assets is in accordance with the control doctrine, but successfully employing such a strategy requires much more than simply acquiring and deploying "control" space assets.

Logistical structure

The logistical challenges of deploying space assets are problems faced by both the US and China, although the launch capabilities of the US (Figure 3.3) dwarf those of the

Chinese. Nevertheless, simply having a launch capability confers a certain degree of technological space control upon the US and China, since, when space assets are brought to bear in future wars, they must be in the environment at the commencement of hostilities. However, while the ability to launch vehicles into space is a key element in exerting space control, an even more important requirement is an on-demand launch system. While neither the US nor China possesses such a capability, such a concept is being considered by the US military. For example, one potential on-demand launch system is the spaceplane, a vehicle capable of high maneuverability, altitude sustainability, and transferring personnel and other critical cargo to low Earth orbit (LEO).

Manned capability

The spaceplane as a key to exerting space control raises the question of what role military personnel will play in space warfare. This is an issue that has not been defined in either US or Chinese space policy, although both countries are developing military manned capabilities. Until the issue of how to use military personnel in space is resolved, the space power doctrine will be supported by a reconnaissance and surveillance capability.

Space surveillance system

Space-based surveillance and reconnaissance assets have global coverage capabilities, providing continuous monitoring of events not just on Earth, but in space also. In addition to providing these capabilities, such a system is essential for the deployment of space weapons such as ASATs, KKVs, and other space control assets such as SBLs.

Space weapons

The issue of space weapons is the subject of Chapter 4, which details the likely anti-spacecraft systems the US and China are likely to deploy in the near future. These assets must not only deny the aggressor the use of low-altitude space, but also be capable of denying the enemy strategic areas of space he wishes to use.

Organizational

How space warfare will be conducted is a problem that cannot be answered until space forces have been developed. It is likely, however, that one function of the space control system will enable military personnel to travel to and be supported in space. In common with terrestrial control doctrines, a space commander will not only

Figure 3.3 Lockheed Martin's *Titan IV* launch vehicle. Courtesy Lockheed Martin (*see colour section*).

command space-based personnel, but also command space control weapons and exert some level of control over a space-based surveillance and reconnaissance system.

SPACE WARFARE

The US, more than any other nation on Earth, relies heavily on the use of space to protect the nation's space capabilities and its investment in space. It is this dependence that drives China's Space Pearl Harbor doctrine, which seeks to strike at US vulnerability in the event of future conflict.[5] Having addressed the subject of what constitutes a space doctrine, it is appropriate to characterize the elements of space warfare before examining the differences between the space doctrines of Washington and Beijing. To that end, what follows is a brief overview of the four national security space missions as they apply to the US and China. Unsurprisingly, the US model encompasses the entire spectrum of capabilities required to exploit space in a manner necessary to advantage its conventional military operations against a range of adversaries. Although China does not yet have all these capabilities, as we shall see, they are capabilities that it fully intends to acquire.

Elements of space warfare

International law has yet to come up with a definition of what constitutes a space weapon. It is perhaps not surprising, therefore, that definitions of "space warfare" are few and far between. In fact, the US military has no definition of space warfare, despite having created the US Strategic Command (USSTRATCOM), an organization dedicated to fighting war in space! Fortunately, Chinese scholars have defined war in space in several different ways, perhaps the most descriptive of which is the PLA's definition:

> "Military confrontation mainly conducted in outer space between two rival parties. It includes offensive and defensive operations between the two parties in outer space as well as offensive and defensive operations between the two parties from outer space to air space or to the ground and vice versa."[6]

While the US may be short on definitions of space warfare, their characterization of military space operations is well defined. What follow are the four elements of space warfare common to both the US and Chinese military.

Space force support

Space force support comprises the two sub-mission areas of launch operations and satellite operations. These sub-missions involve launching satellites into space using expendable launch vehicles (ELVs) and operating those satellites in space (Table

3.1). Typical US launch vehicles used to send large satellites into orbit include the *Titan IV* and the *Atlas V* (Figure 3.4). Once in orbit, Army, Navy and Air Force operators track and monitor the satellites and operate their payloads via a worldwide network of ground stations.

Table 3.1. US space support capabilities.

Launch complexes	*Expendable launch vehicles*	*Reusable launch vehicles*	*Satellite control*
• Western Range (Vandenberg) • Eastern Range (Cape Canaveral) • Launch support and facilities • Range standardization & automation	• Titan 4, 2 • Atlas 3, 2 • Delta 3, 2 • Titan 2 • Taurus • Pegasus • Inertial upper stage • Centaur	• Space Shuttle (due to be retired in 2010)	• US Air Force Satellite Control Network • US Navy Satellite Operations Center

The workhorse of Chinese space force support operations (Table 3.2) is the Long March (LM)-2 launch vehicle (Figure 3.5), which, by the end of 2008, had made 55 launches. Under development is the successor to the LM-2, the powerful LM-5 (see Chapter 7), a next-generation launch vehicle capable of launching 25,000 kg into LEO and 14,000 kg into geostationary transfer orbit (GTO). Due to debut in 2014, the LM-5 will be launched from Wenchang Satellite Launch Center (WSLC), China's new satellite launch facility on Hainan Island.

Table 3.2. Chinese space support capabilities.

Launch complexes	*Expendable launch vehicles*	*Reusable launch vehicles*	*Satellite control*
• Jiuquan Space Facility • Xichang Satellite Launch Center • Taiyuan Satellite Launch Center • Wenchang Satellite Launch Center	• CZ-2 (Long March 2) • CZ-3 (Long March 3) • CZ-4 (Long March 4) • CZ-5 (Long March 5)	• Shenlong (under development) • Space Shuttle (at concept stage)	• Xichang Satellite Launch Center • Taiyuan Satellite Operations Center • Wenchang Satellite Operations Center

Figure 3.4 Lockheed Martin's *Atlas* launch vehicle. Courtesy Lockheed Martin.

Figure 3.5 China's Long March 2-F launch vehicle. Courtesy CNSA.

Space force enhancement

Force enhancement capabilities (Table 3.3) include weather, communications, missile warning, navigation, and various components of signals intelligence (SIGINT). Much of the US force enhancement mission is provided by USSTRATCOM. USSTRATCOM coordinates the use of commercial communications satellites, civil weather satellites, and foreign multispectral satellites in association with programs such as the Defense Support Program (DSP) and the Defense Meteorological Satellite Program (DMSP). Many of these programs were utilized intensively by coalition forces during Operations *Desert Shield* and *Desert Storm*.

Table 3.3. US space force enhancement capabilities.

Reconnaissance	*Surveillance, warning and tracking*	*Weather*	*Communications*	*Navigation*
• IMINT (KH & Lacrosse series) • SIGINT (Vortex 2) • ELINT (Magnum, Orion, Trumpet, Mentor series) • MASINT (Nuclear Detonation Detection System)	• Defense Support Program • Space Surveillance Network	• Defense Meteorological Support Program	• Defense satellite communications system • Milstar • UHF follow-on system	• Global positioning system

Chinese space force enhancement capabilities (Table 3.4), while not as comprehensive as those of the US, nevertheless comprise a diverse set of assets. These assets include an extensive telemetry, tracking and control (TT&C) network, 10 LM variants, and diverse orbital assets ranging from navigation and positioning satellites to surveillance and reconnaissance satellites.

Table 3.4. Chinese space force enhancement capabilities.

Reconnaissance	*Surveillance, warning and tracking*	*Weather*	*Communications*	*Navigation*
• JiangBing 3 (Imaging)	*Satellites* • HaiYang-1 • HaiYang-1B *Control centers* • Xi'an Satellite Control Center[1] • Beijing Aerospace Command & Control Centre[2]	• Feng Yun 1C • Feng Yun 2-06	• DongFangHong 4 (DFH 4) Two operational	• Beidou 1A, 1B, 2A (2), and 5 Five in operation

[1] Xi'an Satellite Control Center, also known as Base 26, is the nerve center of the PRC's TT&C network.

[2] Beijing Aerospace Command & Control Centre (BACC) is the primary command and control centre of key spaceflight missions in the PRC. It is part of Beijing Space City located in northwest Beijing.

Space control

This element of space warfare describes the ability to ensure the freedom of operations within and through the space environment, while denying its use to adversaries. Space control is essential to achieve the force-multiplying effect of all space and missile capabilities. US space control comprises the three sub-elements of space surveillance, National Missile Defense (NMD), and counterspace operations, whereas China's ability to exert space control is limited to space surveillance and counterspace. While many are familiar with the two latter elements, the term *counterspace* is less well known. Simply stated, counterspace operations are those actions required to pre-empt or impede an adversary's access to space assets. This requires various means to target an adversary's space assets, space forces, and third-party space capabilities in an attempt to disrupt, deny, deceive, degrade, and ultimately destroy those capabilities.

To achieve space control, it is necessary to identify what is in orbit, who it belongs to, and what its mission is. In the US, these objectives are accomplished by the Space Surveillance Network (SSN), which tracks, identifies, and catalogs all space objects larger than 10 cm in size (Figure 3.6).

Figure 3.6 Haystack and HAX radars located in Tyngsboro, MA. These radars collect 600 hours of orbital debris data each per year. They are NASA's primary source of data on centimeter-sized orbital debris. Courtesy NASA.

China's space surveillance capabilities are a little more limited than those of the US. An opportunity to observe what many analysts believe was the deployment of a surveillance satellite occurred during China's third manned spaceflight in October 2008, with the release of the BX-1 satellite. Although Chinese officials insisted that the purpose of BX-1 was simply to provide images of the Shenzhou 7 spacecraft, the surveillance capability was subject to close scrutiny when the BX-1 passed within 25 km of the ISS. More worrying are China's counterspace capabilities, as evidenced by its ASAT test – an event that not only outpaced all US intelligence estimates, but signaled China's asymmetric capability of targeting US vulnerabilities.

Space force application

Space force application is a new field of space warfare that includes combat operations in, through, and from space to influence the course and outcome of a conflict. In a future conflict, space force application would be used to attack terrestrial targets from space, thereby minimizing or bypassing high-cost, high-risk conflict. This element of space warfare includes a spectrum of technologies ranging from spaceplanes, reconnaissance satellites, and ASATs, to microwave weapons, KKVs, and orbital bombs. The details of this particular facet of space warfare are discussed in Chapter 4.

US SPACE WARFARE DOCTRINE

In 2002, the US withdrew from the 1972 Anti-Ballistic Missile (ABM) Treaty,[7] a decision causing several nations to call for a ban on weapons in space. While the issue has been on the agenda of the Conference on Disarmament since the mid-1980s, no agreement has been forthcoming, since the Conference requires the consent of all participants and the US not only opposes the action, but has increased its BMD efforts in recent years. US opposition is based on its military space policy that considers that existing arms control agreements adequately protect states and require no further augmentation.[8] The US's attitude to requests for it to desist from pursuing a space weaponization agenda is further illustrated by its non-participation in The Prevention of an Arms Race in Outer Space (PAROS). PAROS is an effort pursued through the UN General Assembly that has repeatedly called for nations to prevent an arms race. Few nations have voted against PAROS, except the US and Israel, who abstain from voting.

US objectives in space warfare

> "We must prepare to face future threats today. My top priority is to ensure Space Superiority. This is at times a difficult concept to comprehend. We did not choose saber rattling words. We selected doctrinal terms; words we know

> are well understood in the Air Force and throughout DOD. The term Space Superiority is akin to Air Superiority. We would not dream of conducting air operations without first establishing Air Superiority. We are not trying to dominate, but we must protect and project our interest in the space medium."
> General Lance Lord, Commander, Air Force Space Command[9]

Given the operational dependence on space, it may surprise some readers that the Department of Defence (DoD) and the intelligence community have yet to develop or agree upon a National Security Space Strategy. In the absence of such a document, some guidance is provided by the Joint Doctrine for Space Operations (JP 3-14) and the US Air Force Counterspace Operations Doctrine (AFDD 2-2.1), which codify US intentions to conduct space warfare. In the absence of a national security strategy, the USSTRATCOM began drafting a National Military Strategy for Space Operations in 2008. This document will provide higher strategic guidance in space to counter emerging threats and ensure the US is capable of achieving space superiority. In the meantime, the US vision for military space dominance is guided by the US Space Command's *Vision for 2020*, published in 1997. The document outlines several US Space Command goals, including the integration of Space Forces "into war-fighting capabilities across the full spectrum of conflict", and to dominate the "space dimension of military operations to protect U.S. interests".[9]

Politics of space superiority

To achieve space dominance, US space strategy is based on two missions. The first of these is *counterspace operations*, comprising "defensive counterspace operations" utilizing ASAT weapons to defend US space assets, and "offensive counterspace operations", which attack enemy satellites. The second mission is *global strike*, which utilizes space-based platforms to attack terrestrial targets anywhere in the world in less than 90 minutes.

Due to the many exotic technologies involved, the extraordinary cost, and the changes in geostrategic outlook that occur with each administration, the US space weapons strategy is still a long way from being implemented. In fact, given these obstacles, it is unlikely the US could realize a space warfare strategy before 2030. However, preparations to achieve space dominance are well underway, as evidenced by several space systems having already made their way into the Pentagon's budget. For example, the first Counter Communications System (CCS) was delivered in October, 2004, with at least two more planned. Another system being funded is the Evolutionary Aerospace Global Laser Engagement (EAGLE) system, a network of space-based laser relay mirrors capable of being used against enemy satellites. Additionally, the XSS-11 experimental satellite, launched in 2005, is capable of autonomously conducting close maneuvers in proximity to orbiting satellites. However, these efforts represent only a fraction of the Pentagon's space warfare projects, most of which are spread across dozens of different accounts, while most of the technology involved is dual-use, meaning it could be used for military or civil purposes. Given that half the military spending on space is classified, it is difficult to even guess what the Pentagon's space warfare-related budget is. Conservative estimates put unclassified and classified DoD spending at $25 billion for 2009, although the real number is probably even higher.

The right to act

Although the deployment of space weapons in space may rub most nations the wrong way, the US pursuit of counterspace capabilities will continue regardless. Part of the reason is that space power is a critical foundation of US military power, which, in turn, is linked to US economic power, and, in turn, to the world's economy. Given these vital bonds, any permanent damage inflicted upon US space systems, such as GPS capabilities, would not only have a damaging effect upon the American economy, but would also deliver a significant blow to the global economy. Another reason is that the US is acutely aware that the Chinese perceive space as America's military Achilles' heel:

> "The United States considers space capabilities – including the ground and space segments and supporting links – vital to its national interest. Consistent with this policy, the United States will: preserve its rights, capabilities, and freedom of action in space; dissuade or deter others from either impeding those rights or developing capabilities intended to do so; take those actions necessary to protect its space capabilities; respond to interference; and deny, if necessary, adversaries the use of space capabilities hostile to U.S. national interests."
>
> US National Space Policy[10]

As stated in the US National Space Policy, America declares the right to act in space while denying other nations that same right. It also reserves the prerogative to develop counterspace measures to prevent other states from hindering US freedom of action in space. While this attitude contravenes the spirit of the Outer Space Treaty (OST, see Table 3.5), it is a posture that the US will continue to pursue, not only to protect its economy, but also to deter China from disrupting or denying US use of space.

Protecting space assets

Since the US relies on space assets for its national security, it must be able to not only assure access to space, but also deny adversaries the use of space. This concept of "space control" does not necessarily require space weapons, since denying the use of space to adversaries is possible through jamming, spoofing, and disabling ground communications links. Also, some proposed space weapons being developed by the US are designed to incapacitate a satellite by simply degrading, denying, or disrupting its signal – a concept known as "tactical denial". One such space weapon is the CCS (Figure 3.7), a ground-based deployable system designed to deny a potential enemy the use of a satellite.[11] Because it is ground-based, the CCS received

Figure 3.7 Northrop Grumman's Counter Communication System. Courtesy Northrop Grumman.

little media coverage when it was declared operational in September, 2004, but it *is* considered a space weapon.

Avoiding arms control limitations

Some may wonder how the US can pursue a policy of deploying space weapons given the arms control limitations and other bilateral agreements restricting the military uses of space (Table 3.5). The answer is surprisingly simple.

While several arms-control limitations call for the "peaceful" use of space, the term "peaceful" remains undefined in the context of international space law. The US persistently defends its position that "peaceful" means "non-aggressive". In other words, any military use of space is in compliance with international space law as long as it does not violate either Article 2(4) of the UN Charter, which prohibits "the threat or use of force", or Article IV of the OST.[15] In fact, US officials have even questioned whether the term "non-aggressive" is too restrictive, arguing there are times when aggression is permissible for the purposes of peace-keeping or self-defense. Using this argument, satellites could be used to support military operations

Table 3.5. Arms control agreements limiting the military use of space.

Limited Test Ban Treaty Prohibits "any nuclear weapon test explosion, or any other nuclear explosion" in the atmosphere, underwater, or in outer space[12]
The Biological and Toxins Convention/Chemical Weapons Convention Prohibits development, production, stockpiling, and acquisition of biological agents, weapons containing toxins, and chemical weapons for hostile purposes[13]
Environmental Modification Convention Prohibits all military or hostile environmental modification techniques that might cause long-lasting, severe, or widespread environmental changes in Earth's atmosphere or outer space[14]
Outer Space Treaty Bars states party to the Treaty from placing nuclear weapons or any other weapons of mass destruction in Earth orbit, installing them on the Moon or any other celestial body, or to otherwise station them in outer space Article IV of the Outer Space Treaty addresses international responsibility, stating that "the activities of non-governmental entities in outer space, including the Moon and other celestial bodies, shall require authorization and continuing supervision by the appropriate State party to the Treaty"

for the purpose of restoring a climate of peace, thereby permitting the use of space weapons!

Ultimately, based on its interpretation of international space law, the US believes there is no legal prohibition on deploying conventional weapons in orbit. Based on this attitude, current US military space policy can be best described as a course of action based on keeping military options open while allowing technology development to drive the evolution of law.

Other limitations

Although international space law places few limitations on the deployment of space weapons and the use of force in space, the US military is still bound by the restrictions of the Rules of Engagement (ROE) – rules under which the US military might fight. For military attacks against the US and for all military operations and contingencies occurring outside the territorial jurisdiction of the US, there exist the Standing ROE (SROE). The SROE are designed to provide guidance for the use of force to accomplish a mission, implement the right of self-defense, and provide rules to apply in peace and armed conflict. The US SROE also include rules applying to military space operations. However, while the SROE ensure that the US military complies with US obligations under domestic and international law, the rules are military directives, *not* law.

Obama's space defense plans

> "It is incumbent on the US armed services to remain open to a wide range of possible capabilities and systems that will enable us to deny our adversaries the advantages gained from space that could be used in a manner hostile to the United States, our citizens, or our national interests. The force structure of the armed services is and will continue to be fully compliant with our international obligations, treaties, and our right to self-defense as spelled out in the UN Charter. If the research and development proves promising and an exhaustive analysis of alternatives concludes that the best/only way to ensure our national security is to base a defensive capability in space, then that option will be provided to the President and Congress for subsequent approval and funding."
>
> Donald H. Rumsfeld, Testimony before the House of Representatives (February 5th, 2002)

Shortly after President Obama took office, a statement appeared on the White House website indicating the administration would seek a worldwide ban on weapons that interfere with military and commercial satellites. While this is clearly a high-priority goal, the chances of it being implemented are remote.

First, before seeking such a ban, the administration must decide on the definition of a space weapon. In 1991, a study conducted by the United Nations Institute for Disarmament Research (UNIDIR) proposed the following definition:

> "A space weapon is a device stationed in outer space or in the Earth environment designed to destroy, damage or otherwise interfere with the normal functioning of an object or being in outer space, or a device stationed in outer space designed to destroy, damage or otherwise interfere with the normal functioning of an object or being in the Earth environment. Any other device with the inherent capability to be used as defined above will be considered as a space weapon."[16]

Unfortunately, if one were to adopt this definition, one could include the Space Shuttle as a space weapon but, at the same time, exclude weapon systems aimed at space, such as the ASAT capability of the Aegis Cruiser (Figure 3.8). In fact, due to the kinetic energies in space and the fragility of satellites, anything with an engine becomes a potential KKV! If the Obama Administration attempted to ban space weapons, how would they account for this dilemma?

One alternative proposed by space-weapons analyst Michael Krepon, of the Stimson Center, is to focus on *capabilities* and not specific weapon systems. By establishing a Code of Conduct for spacefaring nations, Krepon argues, events such as China's ASAT test might be prevented. However, critics of such a code counter that such an agreement would breed complacency among law-abiding nations, while giving rogue nations such as North Korea the advantage of surprise. For example, a rogue nation like Iran, which launched a satellite of its own in February, 2009, would have the ability to attack spacecraft from the ground and create dangerous levels of debris in LEO. The nation most affected by such an act would be the US, with

Figure 3.8 The Aegis Cruiser is fitted with the Standard Missile-3, capable of being used as an anti-satellite weapon. Courtesy Lockheed Martin.

dozens of satellites in LEO, but Iran, with one lone satellite in orbit, would suffer only minimal consequences. It is possible, therefore, that Obama's initiative is aimed at encouraging spacefaring nations to close ranks against such rogue states that present a threat to everyone.

In reality, it would seem that attempts to negotiate a code regulating the behavior of nation-states would be insufficient to preserve security. Such a code would not have prevented the launch of Iran's satellite in February, 2009, it would not have prevented the launch of North Korea's doomed satellite launch in April, 2009, and it would not have deterred China from conducting its ASAT test.

CHINA'S SPACE WARFARE DOCTRINE

Beijing's heavy reliance on secrecy acts in tandem with military deception. It is a strategy designed to limit the transparency in national security decision making, military capabilities, and strategic intentions. Such deception is evidenced by statements opposing the weaponization of space – a stratagem that is practically a staple of Chinese statecraft. Furthermore, the PLA, in keeping with China's obsessive secrecy, does not publish documents akin to the US National Military Strategy, nor does it make current guidelines available for outside scrutiny. Analysis of China's perceptions of thc security environment and the character of modern warfare are therefore based on authoritative speeches and documents, and piecing together editorials from the *Liberation Army Daily*. However, despite Beijing's efforts to conceal and deceive, and the soothing statements of China's civilian scholars opposing the weaponization of space, it was only a matter of time before the red dragon's true space doctrine was revealed when China conducted its ASAT test. In keeping with China's indigenous military tradition, which emphasizes stealth and deception, the ASAT test was in line with China's policy of pursuing counterspace capabilities. Given the potential for China's ASAT test and its emerging counter-space strategy to change the conventional military balance between China and the US, it is worthwhile examining the test in detail before further discussion of Beijing's space doctrine.

Code red? China's anti-satellite test

On January 11th, 2007, China launched an ASAT weapon into LEO (Panel 3.5), destroying an old Chinese weather satellite – a feat demonstrating China's ability, if it so chose, to destroy substantial numbers of US military satellites in similar orbits. A little more than a year later, on February 21st, 2008, the US (which has had an ASAT capability for more than two decades) launched a modified missile-defense interceptor, destroying a US satellite about to make an uncontrolled atmospheric re-entry. However, the US test did not create clouds of dangerous space debris as was the case with China's demonstration.

Panel 3.5. China's ASAT test

On January 11th, 2007, a Chinese medium-range ballistic missile was launched from the Xichang space facility in Sichuan province. The two-stage, solid-fuelled missile, designated the SC-19 by US intelligence, carried a KKV that impacted the *Fengyun-1C*, a Chinese weather satellite orbiting the Earth at an altitude of 864 km and at a velocity of 7.42 km/second. The high-velocity collision generated thousands of fragments, which were ejected at speeds of more than 4,000 km/h into various orbits ranging from 200 km to 3,800 altitudes.[17] Since the test, more than 1,500 objects of traceable debris, measuring at least 10 cm in diameter, have been catalogued and monitored, while NASA's Orbital Debris Program estimated the collision produced more than 35,000 shards larger than 1 cm. In one appalling act of environmental vandalism, China not only created the worst single debris event in the history of LEO operations, but also succeeded in further alienating itself from the international community.

"Put bluntly, Beijing's right hand may not have known what its left hand was doing. The People's Liberation Army (PLA) and its strategic rocket forces most likely proceeded with the ASAT testing program without consulting other parts of the Chinese security and foreign policy bureaucracy – at least not those parts with which most foreigners are familiar. This may be a more troubling prospect than anything the test might have revealed about China's military ambitions or arms control objectives."[18]

Since the Chinese Government initially responded with a mixture of confusion and denial, Western analysts suggested the ASAT test had most likely been conducted without consulting key parts of China's labyrinthine security and foreign policy bureaucracy. Since Chinese officialdom never provided any explanation concerning Beijing's motivation for conducting the 2007 ASAT test, it has been suggested the test was a clumsy attempt to force the US to the negotiating table for a space arms control treaty. Other observers suggested that, with a chaotic year expected in the run-up to the Taiwan elections, it was a somber reminder of Beijing's resolve to defend China's sovereignty at all costs, while others proposed the test was simply a flexing of China's growing military muscle.

In reality, China's ASAT demonstration represented an unambiguous challenge to US dominance in space, triggered by Beijing's observations of US military activities in space that have increasingly shaped China's strategic posture. In 1991, China watched, apprehensively, as the US military demonstrated its satellite communication, reconnaissance, and surveillance capabilities during the Gulf War, a conflict that relied almost exclusively upon space assets.[19] China's observations of the Gulf War, coupled with several key US policy and military documents calling for control of space and the development of space weapons, have caused Beijing to conclude that America is determined to dominate space. The perception of US intent has led Beijing to assume the militarization of space is inevitable.[20] Of more concern to Beijing is the effect US space dominance will have on China's ability to prevail in a conflict in the Taiwan Straits. Unfortunately for Sino–US relations, the US views space dominance as a fundamental tenet of its national security – a stance that breeds a zero-sum competition, in which one side perceives any loss as a gain for the other. Inevitably, such a situation has the potential to not only ignite a possible military confrontation in space, but also to exacerbate the proliferation of space-based weapons.

China's counterspace doctrine

A potential arms race in space is a threat for which no arms-control solution exists. Also, it is a race that is unlikely to be arrested in the near future, since China's leaders feel compelled to respond to American space policies. China's ASAT test represented only the tip of the iceberg of a doctrine designed to counter the overall military capability of the US – a strategy based on the ability to counter American conventional superiority by attacking the vulnerable space-based eyes, ears, and voice of US military power. It is also a strategy designed to defeat superior US conventional forces that China expects to encounter in a war over Taiwan,[21] since it is acknowledged that any orthodox force-on-force encounter, based on attrition, would be doomed from the start. Instead, China is developing indirect approaches to space-based warfare derived from stealth and deception – asymmetric strategies focused on attacking sources of frailty in US capabilities.

Some space analysts may argue that Chinese counterspace investments are simply bargaining chips aimed at negotiating peaceful uses of space with the US. In reality, it is the opposite, as evidenced by Beijing's pursuit of a diverse portfolio of space warfare investments.

Many Chinese officials assume that China is the target of America's missile defense and space planning. Through the eyes of Beijing's politicians, it is inconceivable that Washington would expend such extraordinary resources on the deployment of space weapons for purely defensive reasons. Chinese defense experts fear even a limited space-based defense system could neutralize China's nuclear arsenal, thereby seriously undermining the effectiveness of China's response in any conflict. Beijing's fears are exacerbated by the refusal of the US to declare a no-first-use nuclear policy, a legacy of the Bush Administration's 2001 Nuclear Posture Review (NPR) that specifically mentions the possibility of using nuclear weapons in a conflict in the Taiwan Strait. Perhaps China's greatest fear, however, is the concern that future US space weapons and its missile defense system could subject Beijing to political or strategic blackmail. Such a situation could occur, since space-based weapons would give the US more freedom to intervene in China's affairs, such as their efforts at reunifying Taiwan.

Asymmetric advantage

The only option Beijing has of blunting Washington's vastly superior military space capability is to target the relatively vulnerable communications, intelligence, surveillance and reconnaissance, navigation and guidance capabilities. To that end, China's counterspace doctrine is based on developing asymmetric capabilities based on stealth and deception. Having observed the formidable American military in action during Operations *Desert Storm* and *Iraqi Freedom*, Chinese strategists and planners gained one critical insight convincing them that the US military, while dangerous, is vulnerable and can be defeated by China with the right strategy. The genesis for China's asymmetric strategy is based on observations of the US's aforementioned extraordinary reliance on sophisticated command, control, communications (C3) and intelligence, surveillance and reconnaissance (ISR) systems operating in space. By neutralizing American space-based capabilities, Chinese strategists theorized, it would be possible to give China's military a fighting chance in any future conflict, as suggested by one analyst:

> "An effective defense against a formidable power in space may require China to have an asymmetric capability against the powerful United States. Some have

wondered whether a defensive policy applied to space assets suggests that China's possession of a robust reconnaissance, tracking, and monitoring space system would be sufficient for China to prevent an attack in space and would be in line with China's 'doctrinal' position of 'defensive' capabilities. An effective active defense strategy would include the development of these systems but would also include anti-satellite capabilities and space attack weapon systems if necessary. In essence, China will follow the same principles for space militarization and space weapons as it did with nuclear weapons. That is, it will develop anti-satellite and space weapons capable of effectively taking out an enemy's space system, in order to constitute a reliable and credible defense strategy."[22]

Defending the high ground

Based on its belief that China will be capable of degrading and possibly defeating the superior military capabilities of the US, the Central Military Commission (CMC) of the Chinese Communist Party (CCP) has authorized counterspace programs. These programs are designed to develop direct ascent and co-orbital systems capable of exacting permanent damage on US space-based assets. One of these programs, known as the "Assassin's Mace" program (see Chapter 4), is dedicated to transforming the PLA into a fighting force capable of asymmetrically denying the US's space-based C3 and ISR systems. However, the scope of Beijing's counterspace programs goes beyond achieving basic space-denial capabilities, since China's vision of space warfare involves not just denying space to its adversaries, but using space for affirmative means in a manner mirroring US space doctrine. In fact, the direction in which China's space warfare infrastructure is developing indicates Chinese military space planners seek to deploy space-based combat platforms, achieve terrestrial-strike-from-space capabilities, *and* develop space-based command and control assets.[23] In fact, current Chinese space doctrine is directed at achieving nothing less than the complete spectrum of space combat systems capable of prosecuting the gamut of space support and space attack operations. While many of the programs are protean in their development, China is slowly but surely laying the foundation of a comprehensive portfolio of space warfare systems designed to defend the high ground, which it believes to be the key to any future conflict.

Deciphering intent

In the event of a conflict with China, Washington and Beijing will seek to exercise space control in accordance with the doctrines described in this chapter. Given the limitations of Chinese technological capabilities, the PLA will only be able to exercise such control in a limited area of conflict, meaning the US will be able to prepare the battlefield as it sees fit. However, China's ASAT test sent a clear message to the US that it possesses a devastating asymmetric capability more than capable of countering US space control. China's ASAT test created a lethal cloud of 40,000 space weapons, each capable of inflicting catastrophic damage upon any US satellites that happened to be orbiting at the same altitude. By conducting the ASAT test, China's message to the US was clear: don't count on owning space. With US and Chinese doctrines driven by the need to control space, the potential for LEO and beyond to become a shooting gallery may seem inevitable in a future conflict. But which weapons will be used in such an encounter?

REFERENCES

1. Maj. Gen. Vorobyev. "Strategy", *Voyennaya Mysl* (March–April 1997), 18–24, in FBIS-UMA-97-097-S (April 1, 1997).
2. Lt Col. Drew, D.M. USAF, "Of Trees and Leaves: A New View of Doctrine", *Air University Review*, 40–48 (January–February 1982).
3. Famiglietti, L. "Benign Space Concept Ends with Creation of SPACECOM", *Air Force Times*, 23 (July 12, 1982).
4. "President's Speech on Military Spending and a New Defense", *New York Times*, A-20 (March 24, 1983).
5. Wang Houqing; Zhang Xigye et al. *The Science of Campaigns*. National Defense University Press, Beijing (2000).
6. Hong Bing; Liang Xiaoqui. The Basics of Space Strategic Theory. *China Military Science*, **1**, 23 (2002).
7. *Treaty between the United States of America and the Union of Soviet Socialist Republic on the Limitation of Anti-Ballistic Missile Systems*, 23 UST 3435 (entered into force October 3, 1972, but no longer in effect as of June 13, 2002, due to US withdrawal). Art. XII [ABM Treaty]; US White House, press release, "Statement by the Press Secretary: Announcement of Withdrawal from the ABM Treaty" (December 13, 2001), online internet (January 30, 2005), available from www.whitehouse.gov/news/releases/ 2001/12/20011213-2html.
8. Javits, E.M. "Statement to the Conference on Disarmament", US Mission Geneva, Permanent Representative to CD (February 7, 2002), online internet (January 30, 2005).
9. US Space Command. *Vision for 2020*, p. 3. US Space Command, Peterson Air Force Base, Colorado (1997).
10. US National Space Policy, Office of Science and Technology, www.fas.org/irp/offdocs/nspd/space.html.
11. Uy, H.; Locco, E. "US Air Force Anti-Satellite Weapon Is Operational", www.Bloomberg.com (September 30, 2004).
12. *The Treaty Banning Nuclear Weapon Tests in the Atmosphere, in Outer Space, and Under Water*, 480 UNTS 43 (entered into force October 10, 1963).
13. *Convention on the Prohibition of the Development, Production, and Stockpiling of Bacteriological (Biological) and Toxin Weapons and on their Destruction* (1976), 11 UKTS, Cmd 6397 (entered into force March 26, 1975) [*Biological Weapons Convention*]; *Chemical Weapons Convention 1992*, 32 ILM 800 (entered into force April 29, 1997).

14. *Convention on the Prohibition of Military or any other Hostile Use of Environmental Modification Techniques*, 31 UST 333 (entered into force October 5, 1978).
15. www.oosa.unvienna.org/pdf/publications/STSPACE11E.pdf.
16. Jasani, B. (ed.). *Outer Space A Source of Conflict or Co-operation?*, p. 13. United Nations University Press, Tokyo (1991), published in cooperation with the Stockholm International Peace Research Institute (SIPRI).
17. Covault, C. China's Asat Test Will Intensify U.S.–Chinese Faceoff in Space. *Aviation Week & Space Technology* (January 21, 2007).
18. Gill, B.; Klieber, M. China's Space Odyssey: What the Antisatellite Test Reveals about Decision-Making in Beijing. *Foreign Affairs*, **86**(3), 2–3 (May–June 2007).
19. Wei Chenxi. Space Warfare and War Fighting Environment. *Aerospace China*, Issue No. 10 (October 2001).
20. Tan Xianyu. Study of the Arms and Weaponry of the U.S. Military's Space Warfare in the 21st Century. *Space Electronic Confrontation*, Issue No. 1 (2004).
21. Bush, R.C.; O'Hanlon, M.E. *A War Like No Other: The Truth About China's Challenge to America*. John Wiley & Sons, Hoboken, NJ (2007).
22. Bao Shixiu. Deterrence Revisited: Outer Space. *China Security*, **3**(1), 9 (Winter 2007).
23. Fitzgerald, M. "China's Predictable Space 'Surprise'", *Defense News* (February 12, 2007).

4

Military space assets

SINO–US MILITARY SPACE CAPABILITIES

> "If adversaries are using space in ways that would threaten America or our forces on the battlefield, we have to be able to disrupt or deny their use of those capabilities."
>
> Air Force Lt. General, Michael A. Hamel

The concept of a space weapon has its origins in the works of space visionary, Hermann Oberth. Oberth's publication, *Die Rakete zu den Planetenraümen* (*The Rockets into Interplanetary Space*), mentioned reconnaissance from orbit and suggested the idea of a giant mirror, 100 km in diameter, which could be used to set enemy ammunition dumps on fire! However, while Oberth may be credited with suggesting the notion of space weapons, his writings remained largely undeveloped, and it was left to the Father of American manned spaceflight, Wernher von Braun, to popularize the concept.

The genesis of von Braun's idea of using space weapons occurred a year following his surrender to the Allies in May, 1945, when the US Army asked the great German his opinion of the threat from the Soviet Union. Von Braun responded by suggesting the Soviet threat could only be countered by the development of large, multi-stage rockets as missiles and space boosters. In his vision, orbital nuclear missiles would glide hypersonically half way around the Earth and be guided from a manned orbiting platform 3,700 km ahead of the main station to ensure the missile's impact point was in line of sight.[1] Unsurprisingly, the military deemed the ideas too futuristic, but that didn't stop von Braun. Using the Manhattan Project as an example, von Braun appealed to the Truman Administration for a $4 billion, 10-year commitment to develop this ultimate weapon with the goal of enforcing a *pax Americana* on Earth:

> "We've got mighty little time to lose, for we know that the Soviets are thinking

along the same lines. If we do not wish them to wrest the control of space from us, it's time, and high time we acted."

Wernher von Braun, addressing the Business Advisory Council of the Department of Commerce (September 17, 1952)[1]

Although von Braun's request may have reflected political naïveté, his prediction regarding the Soviet Union was right on the mark. Predictably, von Braun used the launch of Sputnik as an argument to continue the push for weaponizing space. However, after the two superpowers decided it was not in their mutual interest to put bombs in orbit, the nuclear-armed space station was quickly forgotten.

More than 50 years later, von Braun's ideas have gained new relevance as military space advocates assert the need to dominate the High Frontier. While there may be no immediate plans for constructing orbiting nuclear-armed space stations, the space weapons being developed will be capable of inflicting just as much damage. In fact, they may even be used in a manner similar to that envisioned by von Braun more than five decades ago.

The first space war

Before the beginning of Operation *Iraqi Freedom*, the US deployed 6,600 Global Positioning System (GPS) guided munitions, positioned more than 100,000 precision lightweight GPS receivers in Iraq,[2] and used 10 times the satellite capacity utilized during Operation *Desert Storm*. The extraordinary reliance upon space-based assets was further underlined by the fact that more than 100 military satellites supported the US military during *Iraqi Freedom*. The utilization of space-based assets even extended to utilizing the services of the Space Shuttle *Endeavour*, which produced a three-dimensional radar map of targets in Iraq in February, 2000. This massive increase in the use of space technology during *Iraqi Freedom* enabled military responses to occur in minutes rather than hours, resulting in a dramatic reduction in the "kill-chain". It was further proof, if proof was needed, that the High Frontier had become the ultimate military high ground.

Space-based capabilities are becoming ever more integral to the national security operational doctrines of Washington and Beijing. Capabilities such as reliable, real-time bandwidth communications can provide an invaluable combat advantage in terms of clarity of command intentions. Furthermore, satellite-generated knowledge of enemy locations can be exploited by commanders to achieve decisive victories, and precision navigation and weather data from space enable optimal force disposition, decision making, and responsiveness. In short, to implement doctrines aimed at controlling space and denying the use of the space environment to an adversary requires an extensive array of space hardware.

What constitutes a space weapon?

A contentious debate among space analysts has been over what is meant by the term "space weapon" and the degree to which space has already been weaponized. Some observers contend space was weaponized as soon as the Germans launched their V-2 rockets during World War II. However, this assertion conveniently ignores the fact the V-2 rocket made no use of the unique characteristics of space, did not attain orbital velocity, did not release weapons into the new environment, and was intended to accomplish an existing military objective rather than start a space-driven objective. Other analysts maintain space has yet to be weaponized, since no nuclear, kinetic kill vehicle (KKV), or parasitic microsatellites have been deployed in orbit. A third group of experts argue that it is impossible to define a space weapon, since practically any object or vehicle in space could potentially be used as a space weapon. For example, the Space Shuttle (Figure 4.1) could easily be used as a space weapon by simply rendezvousing with a satellite, opening its payload bay, and capturing said satellite! Equally, the kinetic energy of a piece of orbital space debris travelling at more than 7 km/second could easily annihilate any orbiting satellite or spacecraft.

Definition of a "space weapon"

Given the absence of an all-or-nothing definition of what constitutes a space weapon, many officials have used this as an argument to make the case that space

Figure 4.1 Space Shuttle. Courtesy NASA.

arms control is impossible.[3] A universally accepted definition of a space weapon notwithstanding, for the purposes of this chapter, it is useful to adopt the following definition:

> Any system whose use destroys or damages objects in or from space.

Applying this designation excludes non-specific, dual-use systems capable of capturing and disabling satellites as well as missiles passing through the space environment to other targets without damaging space assets. The designation also excludes systems capable of compromising satellite operations (by electronic jamming) and unintentional weapons such as orbital debris in the form of retired satellites. It also excludes space junk such as items lost by astronauts conducting extravehicular activity (EVA). What the designation *does* include is terrestrial and space-based anti-ballistic missiles (ABMs) and anti-satellite (ASAT) systems whose intent is to destroy space assets and other military hardware and/or systems such as nuclear weapons.

While this chapter restricts itself to describing only those weapons that fall into the parameters of the aforementioned designation, it is worth noting that the largest component of a functional space system resides on the ground. These ground-based systems include items ranging from command and control stations and data-handling software to exploitation programs, sensor systems and platforms, and communications links (Table 4.1).

Table 4.1. Spectrum of military space activity.[4]

Space weapons	*Intermediate systems*	*Military operations not involving space weapons*
Key words: *Degrade*, *Destroy*	Key words: *Deny*, *Disrupt*	Communication
Weapons of mass destruction or radiological weapons	ASAT – deny access to satellite or ground system	Navigation
Space-based directed energy weapons	ASAT – temporarily interfere with satellite or ground system (cyber attacks)	Reconnaissance (space-based or high-altitude platforms)
Space-based kinetic weapons such as anti-satellite satellites intended to destroy or degrade other satellites	ASAT – disrupt space operations Ground-based directed (at space) weapons Nuclear weapons for NEO defense	Space-monitoring networks Early warning systems Suborbital delivery of troops or equipment

Space weapon-enabling technologies

The technologies required to design and develop space weapons range from robotics and radiation hardening to artificial intelligence (AI) and advanced cryonics. Since a review of all these technologies is beyond the scope of this chapter, the key enabling technologies have been summarized in Table 4.2. A review of the technologies listed in Table 4.2 highlights the fact that military space technologies are almost identical to those used in the civil space program – a situation presenting a challenging dilemma for the arms control community. Because arms control proponents cannot object to their military applications without also opposing technologies that benefit mankind, existing arms control treaties fail to differentiate between commercial and military space technology. Such a situation has done little to impede the arms race in space.

Table 4.2. Enabling space technologies.

Propulsion/propellants	*Electric power*
Advanced cryogenic	High energy density batteries
Full flow cycle	Lightweight, thermally stable cells
Advanced solid rocket motors	*Structures and materials*
Combined-cycle engines	Lightweight, high-strength composites and ceramics
Electric thrusters	Vibration and thermal control
High-energetic, low-hazard storable propellants	Multi-functional, adaptive structures
"Thinking" satellites	*Synthetic aperture radar*
Autonomous control	Large, light, high-power
Self-assessment	Interferometric
Threat detection	*Signal processors (transmitters/receivers)*
Onboard supercomputing	Higher signal-to-noise ratio
Ground processing	Advanced encryption technologies
Data fusion	*Microelectromechanical systems (MEMS)*
Advanced algorithms for processing and exploitation	Gyroscopes
Antennae	Inertial measurement units
Large, light, controllable	Accelerometers
Higher frequency	Opto-electronics
Steerable beam phased arrays	*Radiation hardening*
Higher-efficiency amplifiers	Techniques and components
	Memory, processors, semiconductor materials

The battlefield in space

Before describing current and future space weapons, it is necessary to characterize the battlefield in space by examining the elements that will be required to fight a

Table 4.3. The battlefield in space.

Mission area	*Scope*
Space transportation	Launch and delivery of payloads to orbit and on-orbit maneuvering
Satellite operations	Control of launch and orbital operations and orbiting spacecraft telemetry, tracking and commanding functions
Positioning, Navigation, and Timing (PNT)	Three-dimensional positioning data and precision timing for users
Command, Control and Communications (C3)	Connection and management of all operational and support missions
Intelligence, Surveillance, and Reconnaissance (ISR)	Data collection from space environment, and processing it into information for use by users/security databases
Environmental monitoring	Observation and knowledge of the space environment
Space control	Freedom of security of space operations; ability to deny use of space to others
Force application	Support from space for defensive/offensive military operations

future conflict. The mission areas listed in Table 4.3 relate to the hardware required to conduct military operations in space and form a useful template for the following section, which compares Chinese and US space capabilities.

OVERVIEW OF SINO–US MILITARY SPACE CAPABILITIES

> "In war, do not launch an ascending attack head-on against the enemy who holds the high ground. Do not engage the enemy when he makes a descending attack from high ground. Lure him to level ground to do battle."
>
> Sun Tzu, Chinese military strategist, The Art of War, circa 500 BC

United States

The Department of Defense (DoD), intelligence community and other civilian organizations (Table 4.4) manage a broad array of military space capabilities. These include launch vehicle development, communications satellites, reconnaissance satellites, and developing systems to protect US satellite systems and deny the use of space to adversaries. The scope of US space hardware ranges from the Aegis Standard Missile-3 (SM-3), capable of intercepting short-to-medium-range unitary and separating ballistic missiles, to multiple kill vehicles (MKVs), capable of delivering several kill vehicles that can attack multiple threat objects within a threat cluster. Supporting the deployment of weapon systems is a complex system of

Table 4.4. Agencies/organizations involved in significant national security space technology activities.

ACDA	Arms Control and Disarmament Agency	**LANL**	Los Alamos National Laboratory
AFA	Air Force Academy	**LLNL**	Lawrence Livermore National Laboratory
AFRL	Air Force Research Laboratory	**MIT**	Massachusetts Institute of Technology
AFSBL	Air Force Space Battle Lab	**NASA**	National Aeronautics and Space Administration
ARL	Army Research Laboratory	**NAVSPACE**	Naval Space Command
ARSPACE	Army Space Command	**NIMA**	National Imagery and Mapping Agency
BMDO	Ballistic Missile Defense Organization	**NIST**	National Institute of Standards and Technology
DARPA	Defense Advanced Research Projects Agency	**NOAA**	National Oceanic and Atmospheric Administration
DCI	Director of Central Intelligence	**NRL**	Naval Research Laboratory
DOA	Department of Agriculture	**NRO**	National Reconnaissance Office
DOC	Department of Commerce	**NSF**	National Science Foundation
DoD	Department of Defense	**NSSA**	National Security Space Architect
DOE	Department of Energy	**ONR**	Office of Naval Research
DOI	Department of the Interior	**OSD**	Office of the Secretary of Defense
DOS	Department of State	**SNL**	Sandia National Laboratory
DOT	Department of Transportation	**USA**	US Army
DTRA	Defense Threat Reduction Agency	**USAF**	US Air Force
FAA	Federal Aviation Administration	**USIA**	US Information Agency
FCC	Federal Communications Commission	**USN**	US Navy
JCS	Joint Chiefs of Staff	**USSPACE COM**	US Space Command
JPL	Jet Propulsion Laboratory		

sensors, including the Space Tracking and Surveillance System (STSS), sea-based radars, and early warning radars. Additionally, STSS includes a command, control, battle management, and communications network presided over by an array of military agencies such as the National Military Command Center (NMCC), USSTRATCOM, and US Central Command (CENTCOM).

Enabling the delivery of much of this hardware is the expendable, medium-lift *Delta II*, one of the US's most dependable launch vehicles, capable of delivering payloads into LEO, polar, and geosynchronous transfer orbit (GTO). Recently, the military has utilized commercial providers such as SpaceX, whose *Falcon I* vehicle is scheduled to launch a TacSat in September, 2009.

In the next decade, the US aims to further utilize commercial providers and improve its space situational awareness (SSA). It will do this by deploying assets capable of providing real-time or near real-time location and status information on spacecraft, in addition to developing systems capable of birth-to-death tracking.

Significant emphasis is also being placed upon protecting US satellites, including stealth and maneuverability defensive options, in addition to working on developing offensive space programs such as kinetic intercept vehicles and ASAT capabilities. However, given the sensitive nature of the weaponization of space, the US is not inclined to say much about these programs!

China

China has been developing its military space assets, including multiple offensive counterspace options, long before it conducted its ASAT test in January, 2007. Beijing's multidimensional space program, also designed to generate the capability to deny others access to space, is reflected in its steady development of military space hardware.

China's launch vehicle capability is centered on 10 Long March (LM) variants, capable of deploying a variety of payloads from LEO to geosynchronous orbits. Recoverable satellites and manned spacecraft are launched from Jiuquan Satellite Launch Center (JSLC) in Gansu Province, while orbital platforms intended for geostationary orbit are launched from Xichang Satellite Launch Center (XSLC) in Sichuan Province. Satellites headed for polar orbit are launched from Taiyuan Satellite Launch Center (TSLC) in Shanxi Province. The Wenchang Space Launch Center (WSLC), currently being constructed on Hainan Island, will be used to launch next-generation launch vehicles to geosynchronous and polar orbits.

Supporting launch operations is an extensive network of ground stations and numerous tracking, telemetry and control (TT&C) facilities spread throughout the country. These facilities are augmented by a fleet of four space event support ships and two other vessels with space tracking capabilities.

Since its first satellite launch in 1970, China has launched dozens of spacecraft. These spacecraft have included communication satellites such as *Chinasat*, *APStar*, and *Sinosat*, and Earth surveillance satellites such as *CBERS-2*, *Haiyang-1*, and *JianBing 5*. China also possesses space-based electronic intelligence (ELINT) and signals intelligence (SIGINT) capabilities in addition to a basic navigation and positioning capability via its Beidou satellite constellation. While the Beidou system is not as accurate as the American GPS system, it could be used to improve weapon accuracy.

In the next decade, China plans to enhance its ELINT and SIGINT capabilities and launch imagery, data relay, and electro-optical satellites. Collectively, these spacecraft will be dedicated to improving the People Liberation Army's (PLA) ability to expand its battlespace awareness and targeting capabilities. In parallel with these activities, China will continue to invest heavily in strengthening connectivity between space systems and military users – a capability that is already secure, survivable, and interoperable among multiple users.

Unsurprisingly, China has made considerable investments in developing counter-space capabilities – a program that has been accelerated since the 1991 Gulf War. China's diverse and comprehensive counterspace program includes upgrading its surveillance and identification systems, developing direct attack weapons such as direct ascent and co-orbital satellites, improving kinetic and non-kinetic means of attack, in addition to exploring directed energy weapons.

CURRENT AND FUTURE CHINESE AND US SPACE WEAPONS

Space transportation

Space transportation includes launch vehicles and propulsion systems for delivering payloads to orbit and capabilities such as on-orbit refueling, servicing, maintenance, and repositioning.

US launch vehicle

The US military utilizes a variety of medium and heavy-lift expendable launch vehicles (ELVs) capable of delivering payloads into LEO, geosynchronous transfer Earth orbits, and all points between. The workhorse launchers of the US space program for the foreseeable future will be the Lockheed Martin *Atlas V* (Figure 4.2) and Boeing *Delta IV* (Figure 4.3) Evolved Expendable Launch Vehicles (EELVs).

Designed to improve access to space by making launch vehicles more affordable and reliable, the US Air Force's (USAF's) EELVs include a standard payload interface and standardized launch pads. The *Delta IV* (Table 4.5) family (Medium, Medium-Plus and Heavy configurations) of launch vehicles is capable of delivering payloads ranging from 4,231 to 12,757 kg to GTO. Central to the *Delta IV* is the commonality between all systems. The Medium and Medium-Plus vehicles use single common core boosters (CCB), while the Heavy variant uses three CCBs. Powering the first stage is a Pratt and Boeing Rocketdyne-built RS-68 liquid hydrogen/liquid oxygen engine, producing 663,000 lb thrust.

Table 4.5. US-evolved expendable launch vehicles.

	Delta IV	*Atlas V*
Manufacturer	Boeing	Lockheed Martin
Height	63–77.2 m	58.3 m
Diameter	5 m	3.81 m
Mass	249,500–733,400 kg	546,700 kg
Stages	2	2
Payload to LEO	3,900–10,843 kg	9,750–29,420 kg
Payload to GTO	N/A	4,950–13,000 kg
Maiden Flight	March 11, 2003	August 21, 2002 (401 variant)[1]

[1] The *Atlas V* launch vehicle system has multiple configuration possibilities, identified using a three-digit naming convention: the first digit identifies the diameter class (in meters) of the payload fairing, the second digit identifies the number of solid rocket motors, and the third digit represents the number of Centaur engines (one or two). Variants to date include: *Atlas V* 401, 501, 511, 521, 531, 541, and 551 configurations.

Figure 4.2 *Atlas V*. Courtesy Lockheed Martin.

Figure 4.3 Boeing's *Delta IV* evolved expendable launch vehicle. Courtesy Boeing.

Lockheed Martin's *Atlas V* uses a single-stage Atlas main engine, the Russian RD-180 engine, and the newly developed CCB TM, with up to five strap-on solid rocket boosters (SRBs). The Centaur upper stage for the *Atlas V* is powered by one or two Pratt and Whitney RL 1-A-4-2 engines, each engine producing 22,300 lb thrust and capable of multiple in-space starts, enabling insertion into low Earth parking orbit or insertion into GTO.

Supporting the *Delta IV* and *Atlas V* launch vehicles is the *Delta II* expendable medium-launch vehicle, managed by the Launch and Range Systems Wing of the Space and Missile Systems Center, Los Angeles Air Force Base. Military payloads are launched off two versions of the *Delta II*. For example, the three-stage *Delta II 7925* (Figure 4.4) has delivered 40 GPS satellites into orbit, whereas the two-stage *Delta II 7920* has launched several imaging and reconnaissance satellites.

A major initiative is the Integrated High Payoff Rocket Propulsion Technology (IHPRPT) program, which coordinates the efforts of the military, NASA, and industry to realize aggressive propulsion technologies. The goals of the IHPRPT program include reduced launch costs, increased satellite life, increased tactical missile effectiveness, and sustainment of strategic systems capability.

Other space transportation technologies being pursued include the X-37B Orbital Test Vehicle (OTV), a winged hypersonic space vehicle that may ultimately lead to a military spaceplane capability. The Boeing Phantom Works vehicle currently being developed is about 9 m long, with a 5-m wingspan. Designed to be launched atop an *Atlas V 501* booster (Figure 4.5), the X-37B (Panel 4.1) is designed for multiple missions. These missions include ISR of ground targets, deployment of micro-satellites for intelligence and surveillance missions, and replenishment of constellations of small satellites that could be carried in the vehicle's payload bay.

Panel 4.1. The X-37B

The X-37B's shape is derived from the winged reusable vehicle concept that gave birth to the Space Shuttle. In common with the Space Shuttle, the X-37B is fitted with thermal protection materials, an experiment bay, and is capable of complex maneuvers such as pitch, roll, and yaw adjustments. Recent testing has been conducted in a series of drop tests from Scaled Composites' WhiteKnightOne carrier aircraft.

Initially, a Space Shuttle "drop test" of the X-37B had been planned for 2006, but was cancelled following the *Columbia* accident. However, the growing military space capability of China and other rogue states has helped to maintain the momentum in the testing and development of the X-37B. Although the X-37B utilizes much Space Shuttle-derived technology, the spaceplane is designed for a quick turnaround of 72 h or less, and will be capable of remaining on station for up to one year.

Figure 4.4. *Delta 7925* launch vehicle carrying the Department of Defense NAVSTAR Global Positioning Satellite GPS-IIR-14. Courtesy Boeing.

Figure 4.5 Artist's rendering of the X-37 orbital test vehicle. Courtesy NASA (*see colour section*).

China's launch vehicles

China's primary launch vehicle is the LM series, of which China has fielded more than a dozen variants, enabling operations from LEO to geosynchronous and polar orbits. The LM-2 is China's largest launch vehicle family, used for LEO and GTO missions. Available in five variants, the LM-2 is a two-stage, liquid-propellant launch vehicle launched from JSLC, and capable of delivering nearly 4 tonnes into LEO. China plans to emulate the US in its quest for a manned spaceplane capability (Panel 4.2).

Satellite operations/Positioning, Navigation, and Timing (PNT)

US military satellites provide meteorological, intelligence, navigation, and communications requirements for the armed services, although meteorological and navigation systems such as GPS are made available for civilian use. In *Desert Storm*, the US relied heavily on military communications systems such as Milstar to communicate from command centers and between troops, early warning satellites to

Panel 4.2. China's Shenlong spaceplane

China's attempt to match the US spaceplane capability is the "Shenlong" (Divine Dragon), an air-launched reusable space vehicle designed exclusively for the purpose of performing military missions.

Due to the obsessive secrecy surrounding Chinese military technology programs, it is not possible to determine whether flight testing of the Shenlong has occurred. However, based on photos of the vehicle attached to the belly of an H-6K bomber and reports that Shenlong will be powered by Russian D-30K turbofans, it is possible to deduce that the spaceplane will be unable to reach sustained low Earth orbit (LEO) flight. Also, since photos of the Shenlong do not reveal indications that it can carry a payload, it is possible the spaceplane is simply designed to test new technologies related to the development of a hypersonic aircraft or possibly a winged space shuttle.

While the latter may seem ambitious for a nation who only achieved its first spacewalk in 2008, the China Academy of Launch Vehicle Technology (CALT) revealed in 2006 that China hopes to have developed a winged space shuttle by 2020.[5]

provide information on missile launches, and military GPS satellites to provide accurate navigation information and guide smart bombs.

The Milstar satellite (Figure 4.6) constellation, consisting of five satellites positioned around the Earth in geosynchronous orbits, comprises the most advanced military communications satellite currently operational. Manufactured by Lockheed Martin at a unit cost of $800 million, each Milstar satellite serves as a switchboard in space by processing information via cross-links with other Milstar satellites and provides voice encryption, data, and facsimile communications.

The US Navstar GPS comprises a constellation of 24 orbiting satellites providing navigation information to both military and civilian users. Operated by the 50th Space Wing, located at Schriever Air Force Base in Colorado, the Navstar (Figure 4.7) system provides extraordinarily accurate three-dimensional location information (latitude, longitude, and altitude), velocity, and time. These capabilities provide the US with a distinct advantage in the strategic high ground – a fact that has not gone unnoticed by China, who is attempting to match Navstar with its budding Beidou GPS system.

In common with the US, China's satellite capabilities support both civilian and military uses. Perhaps China's most sophisticated satellites are its communications satellites such as the DFH-3 and its dedicated military communications satellite, the *Feng Huo-1*. While China is only the third nation to deploy a satellite navigation system, after the US and former Soviet Union, compared with the US's GPS constellation, China's system is, at present, technically inferior. However, China regards its own GPS network as an integral element of its military strategy to not

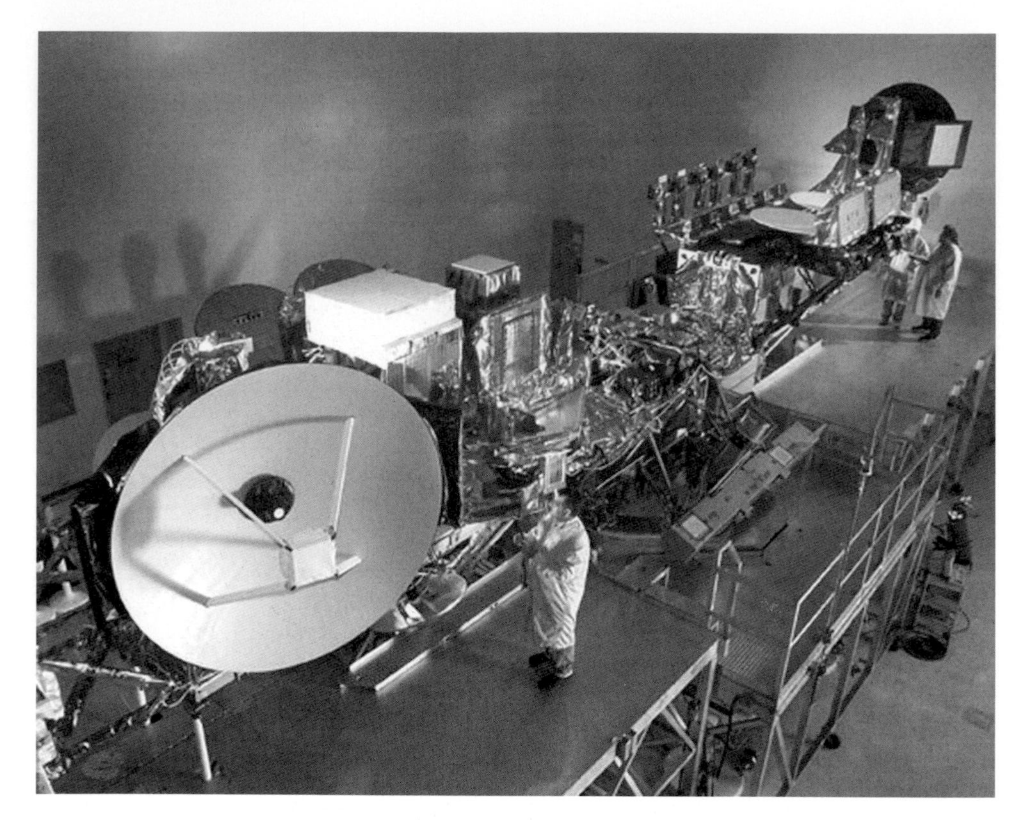

Figure 4.6 Milstar satellite. Courtesy Lockheed Martin.

only safeguard its own territory and assert itself as a regional power, but also to ensure right of access. This latter objective is worth noting, since, with a proven, free US GPS system and Russian and European back-ups in production, the only reason for a Chinese system is supporting offensive military applications.

Command, control and communications (C3)

The US Integrated Space Command and Control (ISC2) program is an element of the USAF's information superiority revolution that provides a flexible platform for emerging military space missions. The ISC2 architecture enables continuous technology insertion and allows interoperability across a range of current and future command and control systems. By providing space-based warfighters with a common operational picture of the space battlefield derived from real-time data, the ISC2 system is a key element in not only providing situational awareness, but also mitigating vulnerabilities that significantly reduce the asymmetric advantage of the Chinese.

China does not have an ISC2 program comparable to the US. Instead, it relies on

Figure 4.7 Navstar satellite in orbit. Courtesy NASA.

a command and control network consisting of the Xi'an Satellite Control Center (XSCC), a number of fixed tracking stations, three mobile command and control stations under the direct command of XSCC, as well as oceangoing instrumentation ships.

Intelligence, surveillance, and reconnaissance

The major feature of the US space surveillance system is the Space Tracking and Surveillance System (STSS), a space-based sensor component of the Ballistic Missile Defense System (BMDS). The STSS will eventually feature a constellation of satellites capable of tracking strategic and tactical missiles and passing missile tracking data to missile defense interceptors.

Beijing's interest in space surveillance capabilities is intimately linked with denying the US targeting data enabling Washington to interdict China's land-based nuclear platforms and elements of its conventional forces. It does not have a system comparable with the STSS, although it has been suggested that the launch of China's BX-1 satellite during the Shenzhou 7 mission in 2008 may be the beginning of a space situational awareness capability.[6]

Environmental monitoring

Worldwide space and terrestrial weather information is provided by the Defense Meteorological Satellite Program (DMSP). The DMSP is managed by the Space and Missile Systems Center (SMSC), Los Angeles Air Force Base, California, together with a team at the National Oceanic and Atmospheric Administration (NOAA), based in Suitland, Maryland. The DMSP satellites, which orbit the Earth at an altitude of 800 km in a near-polar, sun-synchronous orbit (SSO), provide military commanders with space environment data used to assist in high-frequency communications, over-the-horizon radar, and re-entry tasks. Data from these satellites are also used to identify severe weather and to create three-dimensional cloud analyses – information that may affect the launch of space assets.

China launched its first meteorological satellite on May 27th, 2008. The *Fengyun-3A*, which was placed in a polar SSO, carried a suite of instruments, including visible and infrared imagers, an ultraviolet spectroradiometer, and a space environment monitor. Unlike the American DMSP, China's environmental monitoring capability is designed primarily for civilian use.

Space control and force application

Microsatellites

Microsatellite capability is just one of several force application systems in the US's military space arsenal. The Air Force Research Laboratory (AFRL) launched the first experimental microsatellite in January, 2003. The tiny 30-kg Experimental Spacecraft System (XSS-10) was the first in a series of microsatellites ostensibly designed for inspection, rendezvous, and close-up maneuvering around other space assets. However, while the XSS-10's advertised mission is simple rendezvous and inspection activities, by virtue of its maneuverability, the XSS-10 can easily be commanded to ram target satellites, carry explosives, or even be loaded with directed-energy payloads such as high-powered microwave emitters. Following the success of the XSS-10, the AFRL developed the XSS-11, designed to perform on-orbit and beyond on-orbit services including autonomous rendezvous and docking (R&D) with space assets. Given its small size, XSS-11 would be very difficult to detect and, thanks to its proficiency in proximity maneuvering, the XSS-11 (Figure 4.8), just like its predecessor, could easily adjust its speed to ram an adversary's spacecraft.

FALCON

A more overt military space asset is the Defense Advanced Research Project Agency's (DARPA) Force Application and Launch from the Continental United States (FALCON) program. FALCON is a program that is developing a reusable,

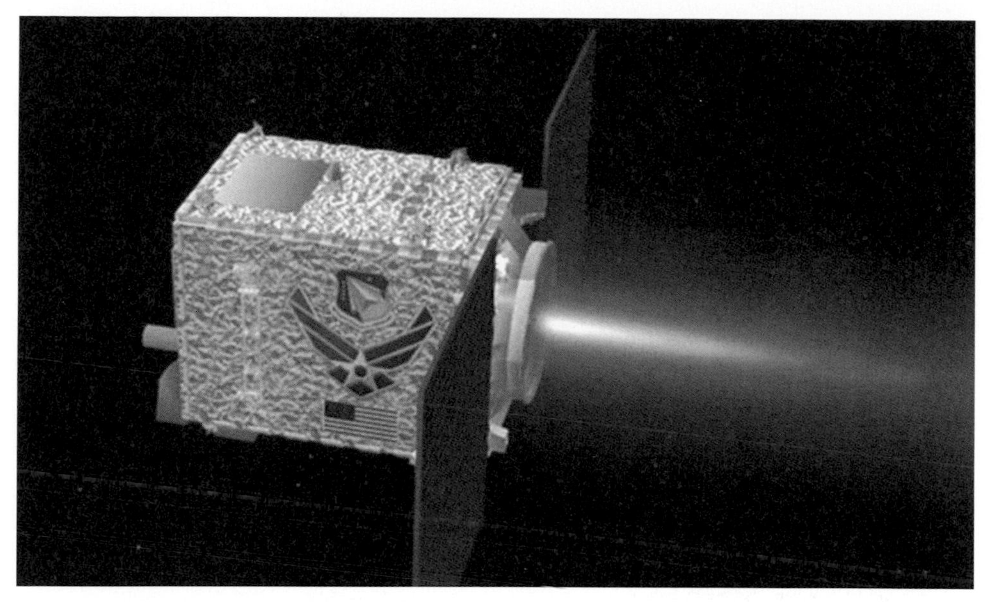

Figure 4.8 XSS-11 satellite. Courtesy Air Force Research Laboratory (*see colour section*).

rapid-strike Hypersonic Cruise Vehicle (HCV) and a Small Launch Vehicle (SLV) capable of accelerating a HCV to cruise speeds, as well as launching small satellites into orbit. In the near term, the FALCON program aims to achieve operational capability for prompt global strike from CONUS by 2010, while a reusable HCV is planned for 2025. Of the two systems, the SLV is most suited to true military space operations, capable of delivering 400-kg payloads to a 28.5° circular orbit at 160 km altitude.

Parasitic satellites

A more current technology is the parasitic satellite, sometimes referred to as Farsat. The Farsat would be launched either as an independent satellite, perhaps by the FALCON SLV, or released by a mother-ship after maintaining a storage orbit different from its intended target. On receiving a command from a terrestrial ground station, the Farsat would activate its ASAT coordinates, before rendezvousing and docking with its target. Given the small size of the Farsats, it is likely that several dozen would be launched months or years before they were required, thereby avoiding a shortage of launch vehicles during the actual conflict.

Kinetic-energy weapons

A technology more radical than the Farsats is the concept of kinetic-energy weapons (KEWs), originally proposed by the RAND Corporation, popularized by science fiction writer, Jerry Pournelle,[7] and currently possibly being developed by the USAF. Perhaps one of the most controversial space weapons, one version of a KEW would comprise hypervelocity tungsten rod bundles being dropped to Earth from orbital platforms. These so-called "Rods from God" would strike terrestrial targets such as hardened bunkers and caches of weapons. While the concept sounds relatively simple, in reality, the laws of physics challenge their feasibility, since there is no way of ensuring the projectiles do not burn up or deform during the high heat generated during re-entry.

Kinetic-energy anti-satellites

In common with China, the US has an ASAT capability based on work conducted by the US Army, which began to develop a direct-energy, kinetic-energy ASAT (KEASAT), capable of being launched by a launch vehicle to destroy hostile spacecraft. Eventually, the DoD transferred control of the KEASAT program to the USAF, but a request for funding the program has not been made for several years. Despite episodic funding, the Pentagon considers the program as completed, although, in the wake of China's ASAT test, it is unlikely the US will conduct a similar demonstration.

Ballistic missile defense system

Perhaps the most widely known space-based weapon system is the Ballistic Missile Defense System (BMDS). The BMDS is a program designed to develop and maintain an operationally effective anti-ballistic missile system that will protect the US against limited ballistic missile threats from rogue nations, nuclear capable states, and against accidental launch of strategic ballistic missiles. To do this, the US is relying heavily on developing space-based assets to enable early detection, launch, and tracking of an enemy ballistic missile, and to engage and destroy ballistic missile warheads above the Earth's atmosphere using kinetic vehicles. The BMDS includes a diverse collection of land, air, sea, and space-based assets founded on cutting-edge and, in some cases, beyond cutting-edge technology. At the sharp end of the BMDS is Raytheon's Exoatmospheric Kill Vehicle (EKV), deployed on top of a long-range ground-based interceptor in Fort Greely, Alaska, and Vandenberg Air Force Base, California. The EKV (Figure 4.9) is really just a euphemism for a space weapon, since space is the only environment in which the EKV will operate (Panel 4.3).

Integral to the BMSD and the employment of the EKVs are the Space-Based Infrared System (SBIRS) and the STSS, which provide rapid early warning and ballistic missile trajectory data. The first phase of SBIRS was declared operational in 2001, and, when fully operational, the system (Figure 4.10) will comprise two payloads in highly elliptical orbit (HEO), four satellites in GEO, and various fixed and mobile ground-based assets. The SBIRS Mission Control Station (MCS), located at Buckley Air Force Base in Aurora, Colorado, not only handles the SBIRS

Figure 4.9 Raytheon's Exoatmospheric Kill Vehicle. Courtesy Missile Defense Agency (*see colour section*).

Panel 4.3. Raytheon's Exoatmospheric Kill Vehicle

The EKV is the intercept component of the Ground-Based Interceptor (GBI) that includes an EELV. The EKV's mission is to engage high-speed ballistic missiles, also known as reentry vehicles, during the midcourse or exoatmospheric phase of their trajectories and destroy them by simply colliding with them. Comprising an infrared seeker and a flight package capable of detecting and discriminating the re-entry vehicle from other objects, the EKV's "hit-to-kill" mission uses only the kinetic energy derived from its velocity to kill vehicles.

satellites, but also centralizes global command, control, and communications for strategic and tactical warning.

Direct ascent and co-orbital ASATs

In addition to its direct ascent weapons such as the DF-31 ICBM, China also plans to develop a co-orbital ASAT interceptor. Launched from Earth into a temporary parking orbit, the co-orbital ASAT would maneuver to attack a specific target, conferring upon China the ability to attack spacecraft whose orbital tracks might not normally traverse the Chinese mainland.

Non-directional attacks

Perhaps the most devastating type of attack is a non-directional attack involving nuclear detonations in space. Such an attack would generate a high-altitude electromagnetic pulse (HEMP) that would degrade and permanently damage satellites' unprotected electrical systems within line of sight (Panel 4.4). In fact, even satellites outside line of sight would be degraded due to the excitation of the Earth's Van Allen radiation belts by the HEMP. While Beijing is technically capable of prosecuting such an indiscriminate attack, it is a route the Chinese would only take *in extremis*, given the dire consequences for its own space-based weapon systems.

The Blackout Bomb

A megaton-class thermonuclear explosion detonated 400 km over Nebraska – a so-called *Blackout Bomb* – would emit a HEMP powerful enough to collapse America's information society from coast to coast in less than a blink of an eye (Panel 4.5).

Figure 4.10 Lockheed Martin technicians prepare the first Space-Based Infrared System (SBIRS) geosynchronous orbit (GEO-1) spacecraft for a major test at the company's facilities in Sunnyvale, California. Courtesy Lockheed Martin.

Panel 4.4. High-altitude electromagnetic pulse

On July 9th, 1962, a 440-km altitude nuclear test known as *Starfish Prime* was conducted by the US military high above Johnston Island in the Pacific Ocean. Unexpectedly, the HEMP emitted by the detonation caused disruptions in electrical systems and equipment in Honolulu, more than 1,100 km away. In addition to shutting down telephone lines, the HEMP disabled three satellites in LEO and the radiation generated by the test eventually destroyed seven satellites by damaging their solar arrays and electronics. More than four decades after *Starfish Prime*, the US is no longer the only nation capable of conducting such tests.

The *Blackout Bomb* is a high-yield nuclear weapon designed to maximize gamma-ray emissions. On detonation, the HEMP moves at the speed of light, striking the Earth to the horizon at line of sight from the detonation. Since such a device would likely be detonated in space, the gamma rays would radiate spherically from the blast point, eventually degrading, and, in most cases, destroying, satellite electronics even at great distances from the blast. A *Blackout Bomb* attack would devastate the electrical grids of the US, wreak havoc on the country's electronic systems, gridlock food, energy, goods and services, cripple the economy, and effectively plunge the US back technologically to the early 20th century. The systems would be down for weeks, months, or perhaps even years. No one really knows.

A future conflict with China may begin with one or more *Blackout Bombs*. One can only imagine the reaction of Chinese warfare strategists when they realized that by detonating one *Blackout Bomb* over the American Midwest, they could throw their arch-enemy back decades in time in industrial capabilities.

To those readers who may think a *Blackout Bomb* attack sounds a little far-fetched, consider the following quote from China's leading authorities on strategy and warfare, who collectively called for the development of weapons that can:

> "Throw the financial systems and army command systems of the hegemonists into chaos. These types of weapons are useful for underdeveloped countries to use against a nation which is 'extremely fragile and vulnerable when it fulfills the process of networking and then relies entirely on electronic computers'. China must abandon the strategy of 'catching up' with more advanced powers and 'proceed from the brand new information warfare and develop our unique technologies and skills, rather than inlay the old framework with new technologies'."
>
> Forum for Experts on Meeting the Challenges of the World Military Revolution, Shijiazhuang, December, 1995

Panel 4.5. How does a HEMP cause so much damage?

As anyone who has seen the film, *The Matrix*, will tell you, an electromagnetic blast can wreak havoc on electronics. In fact, gamma rays from such a blast generate three classes of disruptive electromagnetic pulses, capable of permanently destroying consumer electronics, some car electronics, and large transformers responsible for distributing power. Since EMP is electromagnetic radiation traveling at the speed of light, destruction is instantaneous.

Because of the intense electromagnetic fields generated by a high-altitude detonation, a HEMP would induce large voltages and currents in power lines, communication cables, and other long conductors serving facilities. Examples of typical collectors of a HEMP include railroad tracks, large antennae, pipes, cables, and wiring in buildings. While materials underground would be partially shielded by the ground, they would still be collectors and be capable of delivering the HEMP energy.

An EMP blast occurring at an altitude of 400–500 km results in blast wave interactions between expanding weapon debris and the atmosphere, converting much of the weapon's kinetic yield to ultraviolet (UV) photons, which propagate upward into space. UV photons emitted downward or horizontally are absorbed in the area of the burst point to form a UV fireball. Meanwhile, the flux of damaging X-rays and gamma rays irradiates a considerable region of space, diminished only by spherical divergence.

The size of the hazard zone resulting from a HEMP blast depends on weapon yield, detonation altitude, and the degree of radiation hardening against disruption or harm. The HEMP produces three major energy components that arrive in sequence but which have different effects. The first energy component is the initial energy shockwave, which, despite lasting only one-millionth of a second, is capable of overloading circuitry for every electronic device in line of sight of the burst. The second energy component, which has characteristics similar to a lightning strike, is not as damaging as the first energy component, especially if equipment is fitted with anti-lightning measures. The third energy component is a longer-lasting magnetic signal lasting one-millionth of a second to a full second. This signal inflicts damage primarily upon long-lines electronic equipment such as power and communications transmission lines.

Directed-energy weapons

To those readers who have read the classic science fiction novel, *The War of the Worlds* by H.G. Wells, the concept of directed-energy weapons will be familiar, since the primary offensive weapon used by the Martians is the Heat-Ray. The Heat-Ray is essentially a directed-energy weapon that incinerates anything it comes into contact with and, while the Chinese are not using Martian technology, the characteristics of their directed-energy weapons are similar.

While China's ground-based lasers may not be capable of incinerating people on the spot, they do have the capability of attacking satellites in orbit by dazzling and temporarily blinding the spacecraft by directing sufficient persistent energy in its path. If the energy is particularly high, such a weapon would have the capability to inflict actual structural damage on a spacecraft, causing catastrophic failures to the satellites' power generation and communications systems. While it may sound like science fiction, China has already tested such a system by lasing US reconnaissance satellites.[8]

Assassin's Mace

Perhaps the centerpiece of China's counterspace efforts is the *shashoujian* (Assassin's Mace) program, also referred to as the 998 State Security Project. Established at the behest of former President Jiang Zemin, the purpose of Assassin's Mace is to develop a suite of weapons capable of transforming the PLA into a force capable of winning high-technology wars and to provide asymmetric means by which the weaker Chinese military could defeat the stronger US. The program focuses on a spectrum of capabilities ranging from space technology, cybernetics, and directed-energy systems, to exotic materials and biological warfare. The comprehensiveness of the program has led some observers to suggest the program goes beyond simply seeking space-denial capabilities. In fact, some analysts have declared Assassin's Mace represents a blueprint for achieving space dominance rather than simply attaining counterspace capabilities.[9] Consistent with the expectation that space will simply be treated as another domain in which warfare is permitted, Chinese military writings emphasize that the ultimate high ground must be dominated to secure favorable military outcomes. While most of these writings are aspirational, they nevertheless serve as an indicator of the direction of Beijing's ambitions.

Consequences of US and Chinese space weapons

> "Space superiority is our imperative – it requires the same sense of urgency that we place on gaining and maintaining air superiority over enemy air space in times of conflict."
>
> General Lance W. Lord, Commander, Air Force Space Command

From the US military perspective, the deployment of space weapons deters attacks by reducing the confidence in the success of any attack. By this measure, the more effective the space weapons are, the greater their deterrence value will be, and the less likely the US will need to use them. Deploying advanced military space assets not only confers upon command a better array of options and deters destabilizing operations, such as pre-emptive attacks, but also buys time to understand the strategic consequences and overall impact of military action. While many space doctrine observers and political analysts oppose the weaponizing of space, by implementing a space-based defensive/offensive capability, the US reinforces its strength, which, in turn, enables it to defend its interests and pursue its foreign policy goals. From a US perspective, these rationalizations make perfect sense, since, seen through American eyes, a powerful and influential US is good for maintaining world stability, and enforcing international rule of law. Equally, from the US stance, no nation has the right to deny America their access to space for the purposes of defending the country. However, the positions taken on the deployment of space weapons are not shared by military strategists and politicians in Beijing, who seek to strategically marginalize the US by deploying its own space assets. This threat of marginalization is one that Washington must take seriously, since the US is not so powerful that it can dictate military and political affairs to the world, although some cynical analysts may argue this is the goal of the BMSD!

The case for China

China is a nation in transition and, while many may argue that Sino–US trade relations would deter any conflict between these two space powers, in reality, China's rapid military modernization and its goal of challenging the US for space control are a potentially serious threat. Whether the US and China become adversaries in space may hinge on nothing more than a political change of heart in Washington and Beijing or a confrontation over Taiwan (see Chapter 5).

The case for the US

China's counterspace investments are undeniably diverse, comprehensive, and deadly serious. This should leave no doubt that Beijing is determined to neutralize as far as is possible Washington's operational advantages accruing from the US's overwhelming military dominance. However, many of China's counterspace programs are not mature and others will ultimately be unsuccessful. Furthermore, in the context of the emerging arms race in space, the US's counterspace weapons will increasingly serve as a critical instrument of raising the costs of China's space-denial strategy. This approach will eventually increase the probability that Beijing will desist from asymmetric threats on US space assets.[10] Additionally, as the development and deployment of space weapons move into the future, the US will slowly gain greater and greater control of the space environment. Additionally, countermeasures will become increasingly sophisticated, enabling the US to call upon a global on-call space weapon defense capability and timely response to rapidly evolving threats. Eventually, as interceptor technology becomes more developed, the US will also be capable of countering the HEMP threat and achieving its goal of space superiority.

Arms race inevitability

The US can insist that it means no harm with the development of its space assets, but this will not stop China from taking steps to protect its interests. As China continues to develop its counterspace capabilities, the threat to US military operations will become riskier than ever. However, the arms race in space will not have arisen simply due to a lack of a space arms-control regime, as many left-leaning observers suggest, or even because of the US administrations' disinclination to negotiate an accord that bans weapons in space. The arms race in space will have been entirely rooted in Beijing's doctrine that states China must be able to defeat the US in a regional conflict despite its conventional inferiority. This doctrine of defeating the US asymmetrically has compelled, and will continue to compel, Beijing to exploit every anti-space access and space-denial technology available. While such an arms race is undesirable, it is also unavoidable, since it is grounded in the objective conditions defining the Sino–US relationship – not only one based on divergent political objectives, but one that will persist whether or not the Taiwan issue is resolved. Since the US aims to achieve space dominance, the Chinese will continue to threaten American space assets, forcing the US to adopt an offence–defense arms race, and win.[11]

REFERENCES

1. Neufeld, M.J. "Space Superiority": Wernher von Braun's Campaign for a Nuclear-Armed Space Station, 1946–1956. *Space Policy*, **22**, 52–62 (2006).
2. Lewis, J. *What if Space Were Weaponized.* Center for Defense Information, Washington, DC.
3. "Weaponizing Space: Is Current U.S. Space policy Protecting Our National Security" (May 23, 2007), online at http://nationalsecurity.oversight.house.gov/documents/20070523162935.pdf.
4. Marshall, W.; Whitesides, G.; Schingler, R.; Nilsen, A.; Parkin, K. Space Weapons: The Urgent Debate. *ISYP Journal on Science and World Affairs*, **1**, 19–32 (2005).
5. Coppinger, R. "First RLV by 2020?", *Flight International* (October 17, 2006).
6. Brown, P.J. "China Gets a Jump on US in Space", *Asia Times* online (October 25, 2008).
7. Pournelle, J.E. "Megamissions and Space Power", lecture presented at the United States Air Force War College (March 20, 1994).
8. Muradian, V., "China Tried to Blind U.S. Sats with Laser", *Defense News* (September 28, 2006).
9. www.globalsecurity.org/military/library/report/2008/2008-prc-military-power03.htm.
10. Oberg, J.E. "The U.S. and China: What 'Common Ground' in Outer Space", presentation at the Workshop on Space, Strategy, and China's Future Air War, College Center for Asian Strategic Studies, Keystone, Colorado (June 27, 2006).
11. Tellis, A. China's Military Space Strategy. *Survival*, **49**(3), 41–72 (September 2007).

5

Exploding China's dreams

HOW THE US WILL MAINTAIN SPACE DOMINANCE

> "Our dependence on operations in space, however, makes us somewhat vulnerable to new challenges. It's only logical to conclude that we must be attentive to these vulnerabilities and pay careful attention to protecting and promoting our interest in space."
>
> The Honorable Donald H. Rumsfeld, Secretary of Defense

Many military analysts and space observers assume the term "space superiority" was a recently coined phrase. Historically, however, the expression is credited to Wernher von Braun,[1] who used the term when describing his plans for orbiting nuclear-armed space stations, pre-emptive strikes from orbit, and the development of space-to-ground missiles. Although the focus of the US government shifted away from the concept in the mid-fifties, the "Sputnik shock" briefly resurrected interest in space superiority. In fact, many analysts predicted the Cold War arms race would extend into orbit and perhaps even to the Moon! Eventually, even von Braun conceded the vulnerability of space stations to enemy attack, and the subject of weaponizing space faded.[2]

Forty years later, the Strangelovian concept of space superiority and the desire to dominate the High Frontier with armed satellites and pre-emptive strikes were once again deemed a viable mode of warfare. In 1989, a little known congressional study called "Military Space Forces: The Next 50 Years"[3] described the Pentagon's plans for achieving space dominance. The report was published in book form, and congressional leaders like Senator John Glenn (the former astronaut who was given a Space Shuttle ride in 1999) and Senator Bill Nelson (who got his Shuttle ride in 1986) signed the foreword. In the report, congressional staffer, John Collins noted:

> "Military space forces at the bottom of the Earth's so-called gravity well are poorly positioned to accomplish offensive/defensive/deterrent missions, because great energy is needed to overcome gravity during launch. Forces at

> the top, on a space counterpart of 'high ground,' could initiate action and detect, identify, track, intercept, or otherwise respond more rapidly to attacks."[3]

Perhaps one of the most surprising aspects of the report, given its wild and extreme blueprint for space warfare, was not only that it was commissioned by a Democratic-controlled Congress, but that it was personally endorsed by a group composed mostly of Democrats. The report, reading at times like a good science fiction novel, proposed to Congress that the US needed lunar bases and armed space stations on either side of the Moon's surface:

> "Nature reserves decisive advantage for L4 and L5, two allegedly stable libration points that theoretically could dominate Earth and Moon, because they look down both gravity wells. No other location is equally commanding."[3]

Pursuit of US space dominance

For decades, the US has enjoyed a dominant military advantage in space – a position it has maintained thanks to practically every administration having pursued policies ensuring space dominance. Until recently, US space dominance was assured, since only the wealthiest countries could afford access to space, but, in the past decade, even commercial companies have realized space access capabilities. Since Operation *Desert Storm*, the world's first true space war, the US military has transformed in ways that make space dominance inextricably linked to conducting operations. In fact, the US military has developed such a reliance on space capabilities that President Clinton designated *space* as a vital national interest in his National Security Strategy in 1999.[4]

The Air Force Space Command Strategic Master Plan

The new face of global warfare is outlined in the US Air Force's Strategic Master Plan (SMP). Although parts of it echo some of von Braun's plans, the USAF's agenda is to use space assets to dominate any theatre of war from space. The document details not only how the USAF Space Command is developing exotic weapons capable of destroying any terrestrial target in seconds, but also how the US intends to deny any other nation so much as a foothold in space.

By pursuing an agenda of space domination, the USAF's SMP (Panel 5.1) acknowledges certain policies and international treaties may need to be reviewed and modified. However, it defends this action by boldly stating that the USAF Space Command is the guardian of the High Frontier. While most nations might take issue with the last statement, the US perspective on space superiority shows little regard for the views expressed by other states, since it intends to pursue its space-control directive regardless of any posturing by other nations.

Many defense experts believe the USAF's SMP space directive places the US on a dangerous course, since it clearly rejects any efforts to limit the behaviour of the US military in the space environment. Furthermore, analysts argue, pursuing such an aggressive policy will increase international suspicions that the US is seeking to develop and deploy weapons in space – a concern that possibly precipitated China's ASAT test in January, 2007.

Panel 5.1. Air Force Space Command's Strategic Master Plan

Air Force Space Command's (AFSPC) responsibility is to organize, train and equip space forces by developing, acquiring, operating and sustaining capabilities to exploit and control the space environment. To achieve this, the AFSPC has organized the capabilities into the four mission areas of space force enhancement, counterspace, space force application, and space support.

By addressing these capabilities, the AFSPC aims to dominate the space dimension of military operations across the full spectrum of conflict – a concept known as Full Spectrum Dominance. The synergy of space superiority with land, sea, and air superiority will enable the US military to assure access to space, freedom of operations within the space environment, and a capability to deny other nations the use of space. To that end in the near term (2010–2015), the AFSPC is looking to improve the ability to integrate command and control space forces in any theater of operations. It also seeks to develop technologies to increase standardization of spacecraft design and operations, and develop technologies providing revolutionary capabilities in communications, propulsion, and nuclear strike. In the mid-term (2016–2022), the AFSPC intends to deploy a new generation of space access, global strike, and space superiority capabilities, in addition to finalizing development and deployment of a successor ICBM force. Beyond its mid-term goals, the AFSPC plans to target resources towards deploying space combat forces. This will enable the US military to take the fight to any adversary in, from, and through the space environment. The AFSPC hopes to realize the latter capability in the 2025–2030 timeframe.

Shaping the space environment

The AFSPC roadmap represents a bold plan by the US military to not only shape the military space environment so crucial to US national interests, but also to maintain space superiority, now considered a prerequisite for success in modern warfare. Of course, the de facto space superiority the US enjoyed in Operations *Desert Storm* and *Iraqi Freedom* has not gone unnoticed by China, where government hardliners are planning to match *and* win the battle for space control.

To deter the Chinese, the AFSPC has developed a counterspace roadmap. This roadmap is a time-phased strategy for developing capabilities, enabling the US to subjugate any space capability China might field, while still maintaining American military space superiority. In the event that counterspace capabilities are insufficient to deter an adversary, the US is devising an array of space force application capabilities. These include a Common Aero Vehicle (CAV), capable of dispersing payloads from a suborbital trajectory, and advanced nuclear response capabilities.

Will the ambitious AFSPC SMP succeed? China hopes it will not. However, given the extraordinary levels of US military funding and America's industrial base and technology linkage, the plan seems fiscally possible and technologically achievable. While other states may cry foul, accusing the US of breaking international treaties, in reality, there are no US policies preventing the US from developing or deploying counterspace assets! Furthermore, no international treaties prevent the US from applying force from or through space (see Chapter 3). Meanwhile, the contribution of space systems to global warfare continues to grow. Since other nations lack the financial resources or technology to compete with the US in the pursuit of space dominance, the question is whether America is seizing hold of the future of war.

US asymmetric advantage and vulnerability

Given the huge difference in financial and technological resources between the US and the rest of the world, many analysts believe the rest of the world will simply allow the US to achieve space dominance. These same analysts relegate China to the position of a minor player, unable to compete with the American military juggernaut. However, as evidenced by China's ASAT test, such a view may be a grave miscalculation. The Chinese have long since recognized America's dominance of space as both an advantage and a potential Achilles' heel, as evidenced by statements such as the following from an article in the *Liberation Army Daily*:

> "Currently, space systems have increasingly become systems in which countries' key interests lie. If an anti-satellite weapon destroys a space system in a future war, the destruction will have dealt a blow to the side that owns and uses the space system, stripped it of space supremacy, and weakened its supremacy in conducting information warfare and even its supremacy in the war at large. Anti-satellite weapons that can be developed at low cost and that can strike at the enemy's enormously expensive yet vulnerable space system will become an important option for the majority of medium-sized and small countries with fragile space technology to deter their powerful enemies and protect themselves."[5]

However, the US realizes that its space systems will be an irresistible and tempting target in any future conflict, and is planning accordingly:

> "The American military is built to dominate all phases and mediums of combat. We must acknowledge that our way of war requires superiority in all

mediums of conflict, including space. Thus, we must plan for, and execute to win, space superiority."

General Richard Myers, Chief of Space Command

The US military is a now a system in which information superiority is assured by space-based intelligence, and where orbital communications assets permit ground-based commanders to exert information dominance over adversaries. It is also a system in which global positioning systems (GPS) ensure pinpoint strike capabilities in remote areas and space-based broadband assets provide logistics and surveillance capabilities. However, as the US has continued to pursue unilateral dominance in space, it is has become inordinately dependent on a complex network of space-based intelligence, surveillance, and reconnaissance (ISR) assets and computer-driven command and communications capabilities. While these orbital assets confer the US military a definite asymmetric advantage, they are also the source of acute vulnerability. Given the capabilities of some of the space weapons described in Chapter 4, it doesn't take much imagination to see how some of these weapons might easily be damaged or destroyed, and it is this vulnerability the Chinese counterspace strategy seeks to exploit.

Chinese counterspace reaction

The strategy being developed by AFSPC and America's pre-emptive doctrine is of particular concern to the Chinese. However, while China has demonstrated one level of space-denial with its ASAT test, simply being able to knock spacecraft out of orbit isn't enough. To effectively use its kinetic kill vehicles (KKVs), China is now developing a space awareness capability. Ironically, these resources are available thanks to the vast amount of information on US space systems openly available through international regulatory organizations, and universities! Also, although the ASAT test confirmed that China possesses direct-ascent weapons, these systems represent only one facet of China's counterspace program. Since most of America's remote-sensing, electro-optical, infrared, and radar-intelligence satellites operate in low Earth orbit (LEO), China's direct-ascent weapons would be able to interdict any such satellite in much the same way as the *Fengyun-1C* was destroyed. In fact, direct-ascent vehicles can also be employed to target satellites operating in geosynchronous orbit (GEO),* where US navigation and guidance satellites, military communications platforms and early-warning systems operate. In addition to ballistic missiles, such as the new DF-31, China also has a GEO capability thanks to its current generation of civilian launch vehicles. This capability will shortly be augmented by its development of a co-orbital ASAT interceptor designed to be deployed into a parking orbit, from where it will maneuver to attack its target.

* A geosynchronous orbit is directly above the Earth's equator (0° latitude), with a period equal to the Earth's rotational period The satellite orbits in the direction of the Earth's rotation, at an altitude of 35,786 km.

Once Beijing has acquired an ASAT interceptor capability, it will be able to attack not only spacecraft whose orbital tracks do not traverse mainland China, but also provide a stealth space attack capability. However, stealth ASAT attacks cannot be conducted indiscriminately due to the increasing internationalization of space commerce, which means both sides in a conflict can use third-party satellites. For example, during Operation *Desert Storm* and the Kosovo conflict, the US used European satellites. Therefore, any action taken against a third-party satellite will be viewed as an act of aggression by the country owning it, and the aggressor would have to suffer the consequences of international condemnation and perhaps even military action. Furthermore, China and the US may even share the services of a third-party satellite, so destroying it would compromise China's utilization of space. Exactly how China might employ space assets in a future conflict against the US is uncertain, although it would seem the PLA favor a combination of "soft" and "hard" attacks:

> "Soft kill has the following advantages: broad application, strong operational effect, and does not create pollution in the outer space environment. Still, it cannot cause direct casualties or the destruction of enemy hardware and facilities such as ground launch platforms, space weapons and operation personnel. Using hard destruction can make up for the shortcomings of soft kill methods, containing enemy space capabilities in the long term. Also, hard destruction methods can achieve optimal effect only when complemented with soft kill. Therefore, only by applying in an integrated manner measures of soft kill and hard kill destruction methods can an enemy's space capabilities be weakened or deprived."[6]

As if these emerging capabilities and intentions weren't worrying enough for US military planners, China's pursuit of space dominance will be augmented by its fleets of nano-satellites designed to be launched covertly onboard peaceful space missions. Combined with its direct-ascent weapons, ASAT interceptors, and the development of weapon systems described in Chapter 4, China will gradually acquire a robust counterspace capability that it hopes will eventually lead to its realizing its own goal of space dominance. However, allowing the Chinese even a tiny foothold in space runs counter to the US doctrine of Full Spectrum Dominance, so to counter the Chinese threat, the US will employ counter-counterspace capabilities, guaranteed to ratchet up the arms race in space another notch.

Counterspace consequences

The constellation of space weapons the US intends to deploy in space is geared towards an effective military hegemony designed to deter anyone and everyone from establishing even the flimsiest level of control. US weapon designers are not developing weapons simply to conduct dogfights in space, but to produce sophisticated devices capable of ensuring no nation on Earth dares challenge the US military in space. So, does China really stand a chance?

Consequences of counter-counterspace operations

It would seem that America's vast counter-counterspace capabilities mean China will likely lose any future conflict with the US, and lose badly. However, China's ASAT test is not an anomaly, but an attempt to develop counterspace weapons capable of constraining America's ability to exploit space in a conflict over Taiwan. Gradually, China's counterspace programs will develop and diversify, as Beijing endeavors to negate the operational advantages of Washington's space dominance. Regardless of whether Beijing's counterspace enterprise succeeds or fails, certain consequences are inevitable.

Perhaps the most significant outcome is the death of any agreement banning the deployment, testing, and deployment of space weapons. Given that counterspace operations represent the best chance China has of asymmetrically defeating American military power, there is no way Beijing will agree to space arms control, despite its rhetoric to the contrary. In the absence of such an agreement, Washington and Beijing will be free to embark upon the deployment of weapons in LEO, GEO, and all points in between, in an arms race in space that will put the civilian space race to the Moon in the shade.

A second consequence is the serious threat to American space dominance. Often taken for granted, US space dominance is now threatened by Chinese space-denial programs, exceeding those by Moscow at the peak of the Cold War in both diversity and depth. The US and the Soviet Union were peer competitors, with neither country being hostage to the fears accompanying the power transition that may occur between Beijing and Washington. Such a power transition represents a situation in which China fears being denied the opportunity to secure space domination and the US fears incipient loss of power and influence. To ensure and maintain space dominance, the US will undoubtedly accelerate investment in the areas of systems hardening, autonomous operations, and onboard active defenses. It will also probably build reserve satellites, rapid-response space-launch capabilities, and mobile control stations capable of managing LEO operations in the event of damage to primary control centers. Once again, the consequences of these actions will be an increase in the deployment of weapons in space.

The third consequence is the growth of Chinese space capability as it attempts to ensure deterrence in the event of a conflict with the US over Taiwan. A robust Chinese counterspace capability means such a conflict could result in serious instabilities, perhaps even provoking Beijing to attack the US at the beginning of the conflict, in a "Space Pearl Harbor" approach.[7] Such an attack would inevitably cause the US to retaliate with pre-emptive attacks of its own.

Lastly, the pursuit of Full Spectrum Dominance by the US marks the end of the era of détente. In the first iteration of the arms race, the US, Soviet Union, France, UK, and China adopted the theory of détente as a means of maintaining a nuclear stalemate and to prevent a nuclear exchange. Now, with the demise of the Soviet Union, the US is fashioning its own concept of international relations based on domination and superiority. As the US domination regime gathers speed with the development of ever more sophisticated space weapons and the deployment of space surveillance platforms, the peace-loving states are being left in the dust and the sound of the war drums beats louder.

WINNING AND LOSING A WAR IN SPACE

There are some in the US Defense Department who believe that the more complex the US can make space warfare for the enemy, the more the enemy will be deterred, since, if the enemy has no hope of overcoming the complexities of space warfare, he will be unwilling to initiate war. Given the potential for enemies such as China to engage in asymmetric warfare, such a belief is clearly one that must be revised:

> "China may be poor, but it is a big country nonetheless and possesses satellite reconnaissance capability that even developed countries like Japan, Germany and Britain do not possess. If the Americans tried to interfere with China's internal affairs, such as over the Taiwan question, by military means, they will discover that the Chinese can read their global military moves like the back of their hand. Dealing with China is a lot harder than dealing with most other countries."
>
> Chou Kwan-wu[8]

It is likely that one day, China will attempt to forcibly claim Taiwan, a 395-km-long island lying some 120 km off the southeastern coast of China, knowing full well that such a campaign could lead to a conflict with the US. While most analysts would favor the US in a conflict with China, there are tactical subtleties and attack options that could conceivably change the outcome (Table 5.1). In the following section, a scenario in which China loses a space war with the US is described, before analyzing a different set of parameters in which the tables are turned.

How China would lose a war in space

A conflict between China and the US will almost inevitably be fought over the status of Taiwan, since Beijing has indicated that it will use force if Taiwan takes steps to formalize independence. The following scenario (based on an article by Geoffrey Forden[9]) hypothesizes such a conflict occurring in 2010 and considers current Sino–US space assets and capabilities.

First strike

China's first step in any conflict with the US will be to dramatically compromise the ability of the US to use LEO at a tactical level. To do this, China would first identify LEO satellites passing over Chinese territory on a regular basis. It would then preposition its ASAT-tipped missiles and mobile launchers in remote areas determined by satellite orbits. Simultaneously, China would covertly assemble a

Table 5.1. US space systems and Chinese attack options.

	Communications	*Early warning and nuclear detection*	*ISR*
US space systems	Defense Satellite Communications System (DSCS) Air Force Satellite Communications & Fleet Satellite Communications Military Relay Satellite System	Defense Support Program Space-Based Infrared System-High (GEO) Space-Based Infrared System-Low (LEO)	Electro-Optical Imaging Satellites Infrared Imaging Satellites Synthetic Aperture Radar Imaging Satellites Signals Intelligence Satellites
Chinese attack options (near/medium-term)	Electronic attack Ground attack	Direct-ascent attack Ground attack	Direct-ascent attack Directed energy weapons Ground attack
Chinese attack options (long-term)	Direct-ascent attack Co-orbital attack Directed energy weapons	Direct-ascent attack Co-orbital attack Directed energy weapons	Direct-ascent attack Co-orbital attack

	Meteorology	*Navigation/guidance*	*Remote sensing*
US space systems	Defense Meteorological Satellite Program Geostationary Operational Environmental Satellite	NAVSTAR Global Positioning System	LANDSAT
Chinese attack options (near/medium-term)	Direct-ascent attack Ground attack	Electronic attack Ground attack	Direct-ascent attack Ground attack Directed energy weapons
Chinese attack options (long-term)	Direct-ascent attack Co-orbital attack Directed energy weapons	Direct-ascent attack Co-orbital attack Directed energy weapons	Co-orbital attack

fleet of Long March (LM) launch vehicles capable of intercepting satellites in GEO. Four days before the attack, China would launch the first of its LM rockets, carrying a cluster of ASATs to a parking orbit in LEO. While the first launch would not attract any attention, the next LM launch, following shortly after the first, would set the alarm bells ringing in the Pentagon. Given their current LM inventory and the payload capacity of the LM launch vehicle, it is possible that the Chinese could place 12–16 ASATs in LEO. While the US has dozens of satellites, including 32 GPS satellites, 23 military communications satellites, and more than 90 commercial communications satellites, China would not have to destroy all US space-based assets to compromise US military efforts. If China could destroy those satellites with a direct line of sight of the conflict, it would be able to disrupt US offensive capabilities. However, would China be capable of achieving this? US satellites with direct line of sight would probably include eight military and 22 US civilian communications satellites in GEO. But even if China deployed all its ASATs, it would only be able to destroy 16 satellites, assuming an improbable 100% success rate. Even if China exclusively targeted GPS satellites, the US would still have 16 operating GPS satellites following the attack. However, if the Chinese targeted the nine GPS satellites normally positioned over China, it might be able to eliminate the use of US precision-guided weapons – but only for a few hours, at best. Due to the redundancy in the US space-based satellite network, other GPS satellites would quickly take over the orbits of the damaged or destroyed GPS satellites and the US would be back in business.

Geostationary satellites

If the Chinese decided not to attack the GPS satellites, another likely target might be the five GEO satellites that the US uses to detect missile launches and to alert US nuclear forces. Three GEO satellites simultaneously view the Taiwan area, meaning that the Chinese could destroy all of them, thereby significantly reducing the area covered by US missile defenses against long-range missiles. However, destroying the GEO satellites would place China at enormous risk, since such an attack would lead the US to believe it faced a nuclear attack, thereby resulting in a retaliatory attack in which China would be all but annihilated.

Pre-emption

A more likely scenario is that China fails to damage or destroy any satellites, since the preparations for these attacks would be detected by the US two or three weeks before China assembled its LM rockets on the launch pads. The US would then either initiate hostilities and destroy the rockets before they were launched, or wait until the ASATs were in orbit before shooting them down using its National Missile Defense (NMD) interceptors.

Consequences

Whichever scenario played out, the short-term military consequences of a full-scale attack by China on US space assets are limited at best, even if China achieved a 100% success rate with its ASATs. However, the long-term consequences of an ASAT war for the US and the rest of world would be nothing short of devastating. Assuming the Chinese destroyed nine satellites during the commencement of hostilities, almost 20,000 new pieces of space debris of more than 10 cm in diameter would be generated. Gradually, these fragments would cluster together to form fields of debris, which would render space unusable for hundreds of years or more. Given that the market for GPS receivers alone exceeds $20 billion annually, and the reliance on space assets for forecasting floods and droughts, the creation of such a debris field would cause widespread economic and humanitarian damage.

How China would win a war in space

> "If your enemy is secure at all points, be prepared for him. If he is in superior strength, evade him. If your opponent is temperamental, seek to irritate him. Pretend to be weak, that he may grow arrogant. If he is taking his ease, give him no rest. If his forces are united, separate them. If sovereign and subject are in accord, put division between them. Attack him where he is unprepared, appear where you are not expected."
>
> Sun Tzu, *The Art of War* (c. 500 BC)

A cunning and bold enemy will not face the US's overwhelming military superiority head-on, but will use asymmetric tactics to circumvent American strengths, exploit vulnerabilities, and attack in ways that cannot be matched. Furthermore, counter-space tactics that may be derided by US officials may be deemed perfectly acceptable by a weak adversary attempting to defeat a superpower (the following scenario is based on an article by Brian Weeden[10]).

Pre-emption

While unlikely, it is conceivable that the scenario in which China loses a war with the US could be reversed. On the issue of pre-emption, the US would obviously endeavor to prevent the attacks, but, in reality, the only means to defeat China's kinetic kill ASATs is to launch Trident ballistic missiles (Figure 5.1) – a tactic that may not be effective given the highly mobile nature of the SC-19 booster used in China's inaugural ASAT launch.[11]

Figure 5.1 Trident missile clears the water during the 20th demonstration and shakedown launch from the nuclear-powered strategic missile submarine *USS Mariano G. Vallejo* (SSBN 658). Courtesy US Navy.

Reactive satellite maneuver

China may also be able to overcome the US tactic of changing a satellite's orbital speed – a ploy that, while sound in theory, is much more difficult to achieve in a real tactical situation. To effectively place a satellite out of range, or to change the orbital characteristics of a satellite, it is necessary to conduct either a reactive or a pre-emptive maneuver, neither of which is very effective. To effect a reactive satellite maneuver (RSM) requires a complex set of calculations. These calculations must take into account fuel, velocity change, velocity of the kill vehicle, the velocity change that the kill vehicle can impart to correct its intercept trajectory, and the ability of the seeker head to track the satellite. Since these events require days of planning and cannot be performed autonomously, it is unlikely that the RSM would prove effective, since the time it would take for an ASAT to travel from its launch site to the satellite would be in the order of minutes. Compounding the problem is the ground station network used to transmit instructions to the satellite, which cannot receive instructions until it overflies one of the ground stations. Finally, employing the RSM would mean that the Chinese had already won this stage of the conflict, since, by changing a satellite's orbital characteristics, a satellite would be unable to provide imaging and intelligence information. Of course, there is also the possibility the US could conduct pre-emptive maneuvers of a few satellites, forcing the Chinese to track them, requiring the satellites to overfly Chinese territory at least two or three times. If the satellite continued to maneuver, it is likely the Chinese would be unable to track them, making interception by an ASAT impossible.

The weakness of GPS

One way in which China could save some of its launch vehicles is to selectively deny the US use of its GPS capability by simply targeting the three GPS satellites (theoretically, three GPS satellites are required for a position fix) tasked with providing coverage over Taiwan. In fact, China wouldn't need to use the direct-ascent method to destroy the GPS satellites. Instead, it would employ co-orbital ASATs pre-positioned in orbit months before the attack. The ASATs would then be commanded to maneuver to their targets when required.

Cold reality of space

The US won the Cold War because it was the most militarily powerful nation on Earth. In the same manner in which the Russians challenged the US during the Cold War, Beijing has indicated that it intends to challenge Washington for control of the High Frontier. While the Chinese destroyed one of their own satellites in 2007, the test only replicated a test conducted by the US 22 years previously, with a weapon more advanced than that used by the Chinese. For years, the Chinese blustered about the peaceful uses of space and arms control treaties, before firing a shot across the bow and revealing their true intentions. In reality, China's ASAT test barely constituted a threat to US space dominance because the US military has at least four or five systems (Figure 5.2) capable of damaging and destroying satellites *without* creating an orbital debris cloud. Furthermore, the US, which was technologically superior to the Soviet Union during the Cold War, is, militarily, disproportionately more capable than the Chinese. While Beijing's capacity to develop counterspace and counter-counterspace capabilities will undoubtedly increase, many of the programs it initiates will falter, others will be stillborn, and many will prove unsuccessful. Meanwhile, the US will continue to develop its superior space capabilities and solidify its position of space dominance. The pursuit by the US and China for space superiority will likely continue until China either acquires the capacity to defeat the US, the investments in Chinese counterspace yield diminishing returns, or Sino–American rivalry disappears completely. Since none of these scenarios is likely in the foreseeable future, the US will not waste time and energy attempting to negotiate space-control treaties that are, in any case, unverifiable.

Commanding the future

The Chinese Government may be one of the most corrupt, repressive, and tyrannical regimes on Earth, but it isn't stupid. They are mindful of the lessons of history, where leading powers have been superseded by aspiring nations. They also recognize that those nations that are the most successful in exploiting geopolitical and military options and that act on such options have usually become the dominant powers of the age. Conversely, other states that have consciously failed to act have routinely been relegated to inferior status. Today, the US is faced with consolidating its preeminence in space in a similar fashion to Great Britain's dominance of the oceans in the 19th century. However, with the growing US reliance on space for national security and the inherent vulnerability of space assets, it is almost certain that China will mount a challenge. Inevitably, the US will react to such a confrontation:

> "Given the inherent vulnerability of space-based weapons systems to more cost-effective anti-satellite attacks, China could resort to ASAT weapons as an asymmetrical measure."
>
> Hui Zhang, Chinese space weaponization expert[12]

> "A resourceful enemy will look at our centers of gravity and try to attack them. Our adversaries understand our global dependence on space capabilities, and we must be ready to handle any threat to our space infrastructure."
>
> General Lance Lord[13]

It is possible that America's overreliance on space systems will ultimately prove inconsequential, since the vulnerability of US space assets may not only lead China to attack US space systems, but may also lead them to believe they can conduct a successful military campaign against the US.[14] This bold stance by the Chinese is taken despite knowing that the US considers attacks against its space assets as an act

Figure 5.2 Aegis Standard Missile-3. Courtesy Lockheed Martin.

of war – a position that presents potentially serious challenges to the US military. However, despite the daring posture taken by the Chinese, and despite the seriousness with which China is contemplating space warfare, for Beijing to have any chance of winning a future conflict, it must first gain mastery of space. Given the efforts of the US to improve its space situational awareness (SSA), enhance the survivability of its space platforms, increase autonomous operations, and invest in reserve satellites, it is unlikely that Chinese space dominance will happen any time soon. Furthermore, the US won the Cold War because they were so militarily powerful – a belief that is alive and well today. Since the US intent to ensure its space dominance remains unrivaled, it will move pre-emptively to assert that superiority. This trajectory will inevitably result in the deployment of weapons in space. Similarly, China, whose doctrine also aims to exert military control over the space environment, will, in turn, deploy its military space assets. Inevitably, space will increasingly become an integral part of the US and Chinese military fighting capabilities, not only fuelling the weaponization of space, but serving as the setting for a new arms race.

Onerous responsibility

Is US space dominance inevitable? The idea has been around since the 1940s, but, for most of the Cold War, it remained safely in the realm of science fiction. Not anymore. Every medium – air, land, and sea – has seen conflict, and reality dictates that space will be no different. In preparation for future space conflicts, large-scale wargames have been held regularly since 2001. In the first such space wargame (see Chapter 10), a hypothetical conflict set in the year 2017 involved a space battle with a "near-peer competitor" country (China). During the exercise, the main weapons were laser cannons and microsatellites. Unsurprisingly, the US won, and the message was clear: whoever doesn't control space in the next conflict will lose.

The exercise underlined the reality that space is undeniably the new military frontier. China's ASAT test and the US's SM-3 response are merely ripples in a rapidly unfolding saga, underlining the realization that the world is entering a new strategic era characterized by the weaponization of space. Beijing is challenging Washington's superiority in space because it is Beijing's doctrinal belief that the US is in decline as a space power and that it can, and ultimately will, be replaced by China. The US, recognizing the threat posed by the potential deployment of Chinese space weapons, is moving up a gear on weaponizing space in its efforts to plan, test, and deploy aggressively as the lone superpower. Gradually, the existing military presence in space will proliferate and, after a brief respite from the nuclear competition of the Cold War, the US and China will embark upon a new and costly arms race in space. In so doing, they will assume the onerous responsibility of militarizing space.

REFERENCES

1. Neufeld, M.J. "Space Superiority": Wernher von Braun's Campaign for a Nuclear-Armed Space Station, 1946–1956. *Space Policy*, **22**, 52–62 (2006).
2. von Braun, W. "The Meaning of Space Superiority", ms, c. January, 1958, in USSRC, WvBP, 200-21, published in modified form as "How Satellites Will Change your Life", *This Week* (June 8, 1958), 8–9, 36–37.
3. Collins, J.M. *Military Space Forces: The Next 50 Years.* Future Warfare Series Vol. 4, 1st edn. Brassey's Inc. (November, 1989).
4. National Security Strategy 1999.
5. Li Hechun; Chen Yourong. "Star War: A New Form of War that Might Erupt in the Future", *Liberation Army Daily* (online) (January 17, 2001), 17 (in FBIS as "PLA Article Says Space War May Be Future Form of Warfare", January 17, 2001).
6. www.defensegroupinc.com/cira/pdf/doctrinebook_ch9.pdf.
7. Executive Summary, *Report of the Commission to Assess United States National Security Space Management and Organization*, p. 13. Washington, DC (January 11, 2001).
8. Chou Kwan-wu. "China's Reconnaissance Satellites", Kuang Chiao Ching (in Hong Kong), March 16, 1998, translated in FBIS.
9. Forden, G. "How China Loses the Coming Space War", *Wired Blog Network* (January 10, 2008), http://blog.wired.com/defense/2008/01/inside-the-chin.html.
10. Weeden, B. *China Security*, **4**(1), winter, 137–150 (2008).
11. Lewis, J. "ASATs and Crisis Instability", *ArmsControlWonk.com* (April 15, 2007), www.armscontrolwonk.com/1455/asats-and-crisis-instability.
12. www.spacedaily.com/news/milspace-05zp.html.
13. General Lord, L.W., Commander, Air Force Space Command. "FY06 Defense Authorization Budget Request for Space Activities", Testimony, 10.
14. Wei Qiyong; Qin Zhijin; Liu Erxun. Analysis of the Change of U.S. Military Strategy Before 2020. *Missile and Space Vehicles*, **4**, 1–4 (2002) (in FBIS as "PRC S&T: Analysis of Changing Emphasis in U.S. Military Strategy"). The authors are from the China Academy of Launch Vehicle Technology.

Section III

The Second Space Race

"What American wants to go to bed by the light of a Communist Moon."
Vice President Lyndon Johnson, when asked about
the importance of beating the Soviet Union to the Moon

The prelude to the first space race began in the aftermath of World War II with the work of Wernher von Braun in the US and his colleagues in the Soviet Union. However, except for a few space historians, few people realize that a single man, Lyndon B. Johnson (Figure 6.1), is primarily responsible for not only starting the original space race, but also for ending it.

In 1957 and 1958, Johnson, serving as Senate Majority Leader, created such a storm following the Soviet Union's Sputnik launch that Eisenhower was forced into

Figure 6.1 Spiro Agnew and Lyndon Johnson watch the *Apollo 11* lift off. Courtesy NASA.

a space race he didn't want. Nearly a decade later, Johnson, now President, was forced to almost shut down the program he had worked so hard to sell due to the expenditures required to pay for the escalating Vietnam War. In 1966, with war costs spiraling out of control, President Johnson's Administration was forced to consider the daunting prospect of having to cut space spending to pay for Vietnam – a decision that would hand victory in the space race to the Soviet Union.* In an effort to prevent the Soviet Union from using victory in the space race as a means to gain control of space, Johnson offered the Soviets a deal for mutual renunciation of the prizes available by negotiating the Outer Space Treaty (OST). The OST[1] essentially obliged the two superpowers to agree to no nuclear weapons in space and an understanding that no country could lay claim to the Moon. To Johnson's surprise, Russian leader, Nikita Khrushchev, agreed, and the OST was ratified in 1967. Although President Johnson managed to preserve the Moon landing, space spending decreased shortly after Apollo (Figure 6.2) and, for the next four decades, space development has been restricted to low Earth orbit (LEO). This stagnation of space exploration has caused more than one space analyst to suggest ways to restart a race to space in the hope of reinvigorating the spirit of exploration. With the arrival of China in the manned spaceflight arena, the prayers of those analysts have finally been answered, although perhaps not under the circumstances they had hoped.

The US and China have announced intentions of landing humans on the Moon by 2020, and are now engaged in the early stages of a new space race. However, the new race appears to have few of the hallmarks of the one between the US and the Soviet Union, when space was a surrogate battleground for geopolitical supremacy. While the adversary in the new space race is a Communist nation and the new space race juxtaposes contrasting political ideologies in a situation paralleling the Communist Soviet Union regime pitching against US democracy, the similarities end there. Whereas money was not an issue for the Apollo program, which received more than 5% of the federal budget, NASA now receives a little over half of 1% of the budget – a situation that has some analysts suggesting that the program isn't funded to succeed. Also, whereas the first space race was characterized by overwhelming public support, NASA's new program is viewed with skepticism and disinterest. Frankly, many just don't care. Meanwhile, America's competitor has the ability to impose its political will in a system that doesn't need to fund Social Security and whose government isn't accountable to a democratically elected Congress. Given these circumstances, it isn't surprising some have suggested that American astronauts might find a welcoming committee on the lunar surface when they return! Against this background, it is useful to remind ourselves of the history of the first space race.

Who won the first space race and when was it won? Some may argue the US won

---

* At the time, Sergey Korolev, the Soviet's genius rocket designer, who was the key to the Soviets' beating the US to the Moon, died during colon surgery. However, the US did not know at the time, and President Johnson's intelligence services were telling him that the Soviets were on course to reach the Moon.

Figure 6.2 *Apollo 8* launch. Courtesy NASA.

Figure 6.3 *Apollo 8* crew. Courtesy NASA.

the moment Neil Armstrong set foot on the Moon, while others would argue the Soviet Union won when they launched Sputnik. Most historians agree the first space race began with the launch of the Soviet Union's first satellite, *Sputnik I*, on October 4th, 1957 – an event that sent the US into a frenzy. In the struggle to catch up, President Eisenhower created the National Aeronautics and Space Administration (NASA) on July 29th, 1958. On May 25th, 1961, just six weeks after Yuri Gagarin's inaugural trip into space, the finish line of the space race was announced by President Kennedy. By promising the world that the US would land astronauts on the Moon before 1970, Kennedy raised the bar in the space rivalry stakes by taking aim at a goal that would put the achievements of the Soviet Union in the shade. As the space race progressed during the 1960s, the Americans struggled through setback after setback, as the Soviet Union achieved back-to-back accomplishments such as the first long-duration flight in June, 1963, the first woman in space, also in June, 1963, and the first extravehicular activity (EVA) in March, 1965. However, in a dramatic comeback, which started in 1968 (Figure 6.3), the US snatched a decisive victory on July 20th, 1969, when Neil Armstrong became the first person to step on the Moon, thereby ending the first space race that had cost the US $25 billion.

Since that momentous moment in space history, the space race fervor subsided, and both the US and the Soviet Union became stuck in LEO. For decades, the incremental future of humans in space was restricted to sending Space Shuttles into LEO, building space stations, and servicing the Hubble Space Telescope. But, in January, 2004, President Bush announced to the world that he was putting NASA on

Figure 1.4

Figure 2.1

Figure 2.5

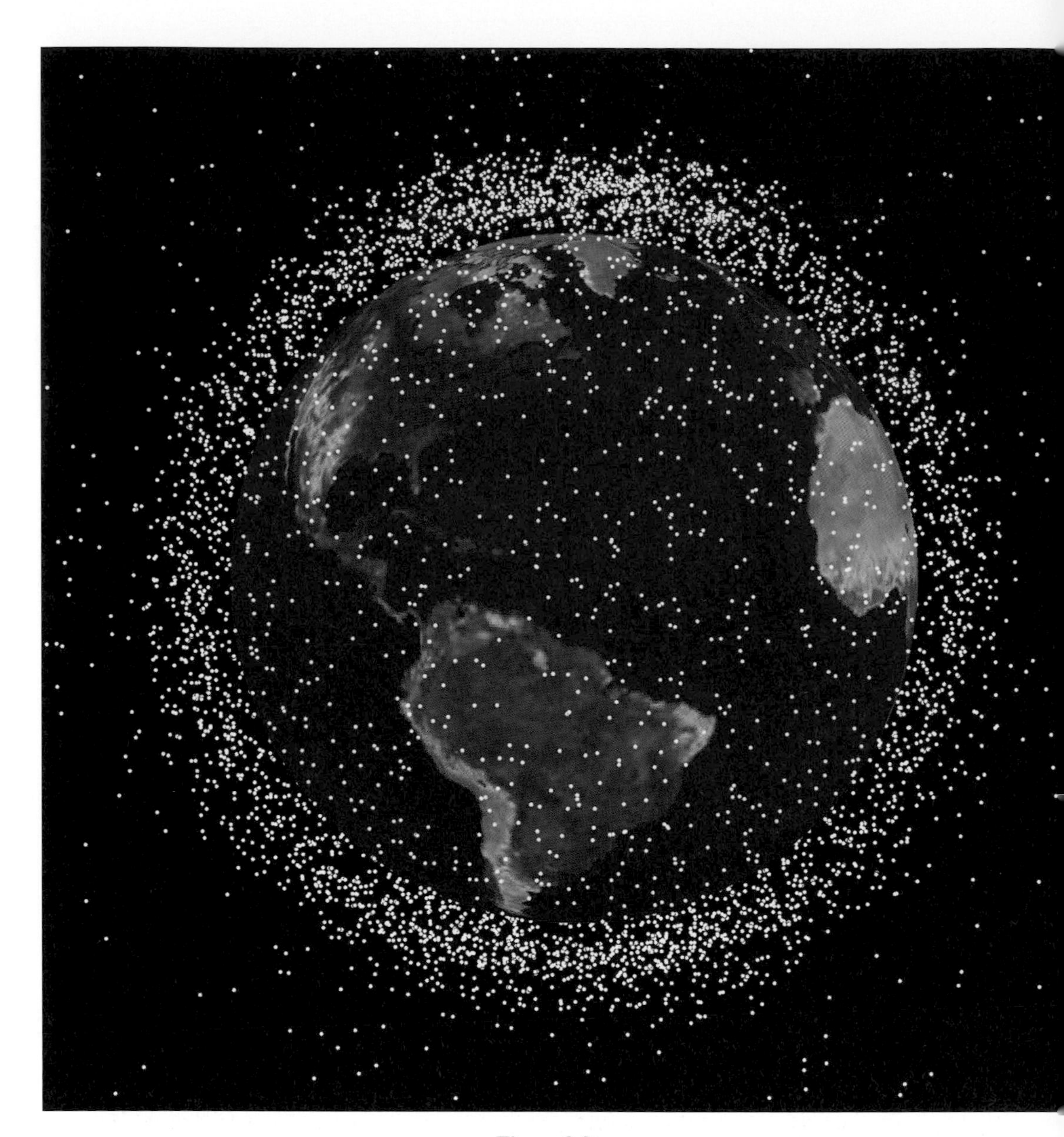

Figure 2.8

Figure 3.1

Figure 3.9

Figure 4.5

Figure 4.8

Figure 4.9

Figure 6.4

Figure 7.5

Figure 7.7

Figure 7.11

Figure 7.12

Figure 8.1

Figure 8.4

track to return to the Moon by 2020, and to embark upon manned missions to Mars and beyond. The following month, not to be outdone, the Chinese Space Program announced that it, too, would send taikonauts to the Moon by 2020, and so a new space race was born. Already, the first salvos of the competition are being fired, with China announcing its plans to launch a space station and how it intends to mine the Moon of Helium-3. Here, in Section 3, we shall see that, like the first space race, the new space race is taking place in the context of dramatically different resources and international prestige. While NASA's annual space budget exceeds $19 billion, the Chinese are attempting to compete with a budget that is barely $2 billion a year. We will also see how China's successes in space may reenergize NASA, an agency that has been in a state of torpor since the *Columbia* tragedy, and why the objectives of this new competition are not about "flags and footprints", but about leadership in space.

6

Chinese and American space exploration programs

In January, 2004, the Bush Administration launched the Vision for Space Exploration (VSE), a bold and forward-thinking space exploration policy, directing NASA's human spaceflight program for decades to come. To accomplish the VSE,[2] NASA initiated the Constellation Program, tasked with developing a Crew Exploration Vehicle (CEV) and a new heavy lift cargo launch vehicle (CaLV). Unfortunately, in common with previous presidents, Bush seemed to forget his enthusiasm for the US space program almost immediately after vacating the announcement podium! The lack of follow-up to the VSE echoed the forgotten policies of President H.W. Bush, who, although more committed to the space exploration program than his son, was unable to ensure the survival of the Space Exploration Initiative (SEI), the program created during his tenure.

THE VISION FOR SPACE EXPLORATION

NASA's plan (Table 6.1) for exploring space and sending astronauts to the Moon and beyond is both enterprising and far-reaching. In fact, the stated goal of conducting manned missions to Mars has caused some critics to suggest that the VSE may be too ambitious.

Table 6.1. Overview of NASA's Constellation Program.

Complete the International Space Station	Extend sustained human presence to the Moon, to enable eventual settlement
Safely fly the Space Shuttle until 2010	Strengthen existing and create new global partnerships
Develop and fly *Orion* no later than 2015	Develop supporting innovative technologies, knowledge, and infrastructures
Return to the Moon no later than 2020	Extend human presence across the solar system and beyond
Use the Moon to prepare for future human missions to Mars and other destinations	Expand Earth's economic sphere to encompass the Moon and pursue lunar activities with direct benefits to life on Earth

Constellation overview

Hardware

The Constellation Program includes new space hardware designed to support operations to LEO, Moon missions, and, eventually, manned missions to Mars and beyond. To achieve these objectives, Constellation is building two new launch vehicles, two manned spacecraft, and a variety of surface support equipment for exploration and scientific research. The first of the new launch vehicles, scheduled for its first unmanned suborbital test flight in October, 2009, is the two-stage *Ares I* (Figure 6.4), designated the Crew Launch Vehicle (CLV). Based on Shuttle technology, *Ares I* will rely on an extended Solid Rocket Booster (SRB) to launch astronauts into orbit. Once the proving flights of *Ares I* are over, NASA will begin to prepare for returning astronauts to the Moon, which will require the services of *Ares V* (Figure 6.5), the heavy lift CaLV designed to carry lunar hardware such as the Earth Departure Stage (EDS) and the lunar lander into LEO.

The *Ares I* and *Ares V* designs incorporate heritage elements such as the modified SRBs and External Tank (ET), whereas the CEV, known as *Orion*, is a new man-rated vehicle capable of operating autonomously for up to 210 days. Capable of

Figure 6.4 Artist's rendering of *Ares I* on the launch pad. Courtesy NASA (*see colour section*).

Figure 6.5 Artist's rendering of *Ares V* launch. Courtesy NASA.

accommodating up to four crewmembers for lunar missions and six crewmembers for Mars and International Space Station (ISS) missions, *Orion* also has the capability to deliver pressurized and unpressurized cargo to the ISS.

The Lunar Lander, known as *Altair* (Figure 6.6), is the second man-rated vehicle of the Constellation Program. Capable of transporting four crewmembers to the lunar surface and providing global access capability, *Altair* will also provide crewmembers with a return-to-Earth capability at any time during the mission.

Realizing the VSE: commissioning the Exploration Systems Architecture Study

The month following President Bush's address, NASA Administrator, Sean O'Keefe, released the VSE, NASA's roadmap of the agency's many programs that support the vision laid out in the speech. To develop the many technologies and systems required to realize the VSE, the Exploration Systems Mission Directorate (ESMD) was created. The following year, in April, 2005, NASA's new Administrator, Dr Michael Griffin, commissioned the Exploration Systems Architecture Study (ESAS), which evaluated ways in which NASA could realize the VSE. The ESAS represents one of the primary building blocks of the Constellation Program, as it documents how the VSE will be achieved.

Figure 6.6 Artist's rendering of *Altair* on the Moon. Courtesy NASA.

EXPLORATION SYSTEMS ARCHITECTURE STUDY CHARTER

The ESAS began on May 2nd, 2005, and comprised 20 core team members, collocated at NASA HQ, and hundreds of NASA employees from various NASA centers. To direct the ESAS team, a number of Ground Rules and Assumptions (GR&As) were established, which included guidelines and constraints directed at factors such as safety, operations, technical, cost, schedule, testing, and foreign assets.

Exploration Systems Architecture Study Ground Rules and Assumptions

The Safety and Mission Assurance GR&As guiding the ESAS team was NASA's Procedural Requirements (NPR) document known as the Human-Rating Requirements for Space Systems. This document comprised a set of requirements used as a reference when the ESAS team defined mission aspects such as abort opportunities and orbital operations.

Notable *Operations* GR&As included requirements to deliver crew and cargo to and from the ISS until 2016, to devise an architecture separating crew and large cargo to the maximum extent possible and for all on-orbit flight operations to be conducted at the Kennedy Space Centre (KSC). Examples of some of the *Technical* GR&As included restrictions requiring *Orion* to be designed for a crew of four for lunar missions and a crew of six for ISS missions. The ESAS team was also required to follow certain factors of safety (FoS) for elements such as the crew cabin and FoS margins for rendezvous and docking (R&D) operations. One of the *Schedule* GR&As was a requirement for the first *Orion* flight to the ISS by 2011 and to perform the first lunar landing by 2020. However, since the publication of the ESAS study, the first of these requirements has slipped to 2015 and it is uncertain how much this slip will impact the lunar landing date. Another important guideline is the *Testing* GR&As, which require at least three fully functional flights be completed prior to *Ares I* being human-rated, and one flight test of *Ares V* before it can fly high-value cargo.

Exploration Systems Architecture Study tasks

Some of the ESAS team's first tasks included assessing *Orion*'s requirements, defining how *Orion* could provide crew transport to the ISS, and devising a plan to reduce the time between the Shuttle's retirement in 2010 and *Orion*'s first manned flights. Next, the ESAS defined the requirements and configurations for the launch vehicle systems to support manned missions to the Moon and Mars. The ESAS then developed reference lunar architectures for the manned missions to the Moon and identified the technologies required to enable these architectures. These four tasks were transposed into the four major points of the ESAS Charter:

(1) Definition of *Orion*,
(2) Definition of *Ares I* and *Ares V*,
(3) Definition of the lunar architecture, and
(4) Definition of the technology required.

To fulfill their tasks, the ESAS effort performed hundreds of trade studies examining options ranging from an assessment of different *Orion* shapes to determining optimum EVA requirements. These trade studies also quantitatively and qualitatively assessed costs, schedule, reliability, safety, and risk associated with all factors related to the elements of the architecture required to return humans to the lunar surface.

Mission architecture overview

Once the ESAS team had performed the trade studies, they devised a series of Design Reference Missions (DRMs). The DRMs defined how crews and pressurized and unpressurized cargo would be transported to and from the ISS. The DRMs also defined how crew and cargo would be transported to and from the lunar surface and how crew and cargo would be transported to and from an outpost at the lunar South Pole.

Design Reference Missions overview

DRM: transportation of astronauts to and from the ISS

The objective of this DRM is to ferry astronauts to the ISS, where they will complete six-month mission increments. During the mission, crewmembers will be provided with an any-time return capability made available by *Orion* berthed at the ISS. *Orion* will be launched by *Ares I* into LEO, after which a series of burns raise the orbit of *Orion* so that it eventually catches up with the ISS. Once *Orion* is close to the ISS, a standard R&D maneuver will be performed, the crew will ingress, and *Orion* will assume the role of "rescue vehicle" for the duration of the mission. Upon completion of the mission, *Orion* will execute a de-orbit burn, and perform a splashdown off the California coast.

DRM: transportation of unpressurized cargo to the ISS

The mission's objective will be simply to haul unpressurized cargo to the ISS, using a Cargo Delivery Vehicle (CDV) and *Ares I*. Following launch into LEO, the CDV will perform a series of orbit-raising burns to close with the ISS before conducting a standard onboard-guided approach designed to place it within reach of the ISS's Remote Manipulator System (RMS). The ISS crew will use the RMS to grapple the CDV and berth it to the ISS, where it will remain for 30 days before it will be unberthed and conduct a ground-validated de-orbit burn for a destructive re-entry.

DRM: Transportation of pressurized cargo to and from the ISS

The purpose of this mission will be to ship pressurized cargo to and from the ISS and return to Earth following a 90-day berthing period at the ISS. The mission profile will utilize a cargo variant of *Orion* (CVO), loaded with up to 3,500 kg of logistics, atop *Ares I*. Following LEO insertion, the CVO will perform orbit-raising burns to chase the ISS. Once at a safe station-keeping position relative to the ISS, Mission Control will command the CVO to perform an onboard-guided approach, the CVO will dock, pressurization checks will be conducted, and ingress of cargo will be performed by the ISS crew. Following a 90-day docking phase, the CVO will perform a de-orbit burn and perform a splashdown off the California coast.

DRM: manned lunar mission with cargo

The lunar sortie mission is designed to transport four astronauts to any site on the Moon – a capability referred to as *global access*. Once there, the astronauts will spend up to seven days performing scientific and exploration tasks in pairs. A lunar sortie mission will require *Ares I*, *Ares V*, *Orion*, a Lunar Surface Access Module (LSAM), and an EDS. The mission will commence with the launch of the *Ares V*, which will deliver the LSAM and EDS to LEO. *Ares I* will then deliver *Orion* to LEO, where *Orion* will R&D with the LSAM–EDS configuration. The EDS will perform a Trans-Lunar Insertion (TLI) burn and be discarded. The LSAM will then perform a burn inserting the LSAM and *Orion* into lunar orbit. Once in lunar orbit, the crew will transfer to the LSAM, undock from *Orion*, and descend to the lunar surface. After spending seven days exploring the lunar surface, the LSAM will perform an ascent to low lunar orbit (LLO) and R&D with *Orion*, which will return to Earth and splashdown off the California coast.

DRM: cargo transportation to lunar surface

This mission will help to establish a permanent human presence on the Moon by ferrying up to 20 tonnes of cargo to the lunar surface. To achieve this objective, the *Ares V* will launch an EDS and cargo variant of the LSAM into LEO, the EDS will perform the TLI burn, and the LSAM will perform the burn to insert itself into lunar orbit.

DRM: outpost mission with crew and cargo

To support a six-month surface increment on the lunar surface, this DRM will transfer a crew of four astronauts and supplies in a single mission using a suite of vehicles. First, the LSAM and EDS will be deployed following a single *Ares V* launch into LEO. Then, *Ares I* will launch *Orion* into LEO and will R&D with the LSAM–EDS configuration. After performing the TLI burn, the EDS will be discarded and the LSAM will perform the lunar orbit insertion (LOI) burn for *Orion* and the LSAM. Once in lunar orbit, the crew will transfer to the LSAM, descend to the lunar surface, and commence a six-month increment on the surface. At the end of the mission, the LSAM ascent stage will return the crew to lunar orbit, where the LSAM will dock with *Orion* and the crew will transfer to *Orion* for the journey back to Earth, leaving the LSAM to impact the lunar surface.

The DRMs devised by the ESAS represent a time-phased evolutionary architecture designed to return humans to the surface of the Moon. The details of the DRMs and other elements of the architecture such as the design of *Orion* and *Ares I* were published in the ESAS report and, shortly after the report's publication, on December 30th, 2005, NASA's plans for realizing the vision became law with the passing of the NASA Authorization Act of 2005:[3]

> "The Administrator shall establish a program to develop a sustained human presence on the moon, including a robust precursor program to promote exploration, science, commerce and U.S. pre-eminence in space, and as a steeping stone to future exploration of Mars and other destinations."
>
> NASA Authorization Act of 2005

THE CONSTELLATION PROGRAM

To achieve the VSE's goals, the Constellation Program is leveraging expertise across NASA – a strategy that is gradually transforming the centers into a unified agency-wide team. Although the continued operations of the Space Shuttle constrain the budget and workforce available to develop Constellation through 2010, NASA is making the best use of its experienced workforce as they become available through the Shuttle's phased retirement. Already, targeted activities in the lunar capability development have begun, initial concept development of the lunar lander and *Ares V* are underway, and a schedule of the major vehicle engine tests and missions has been published (Table 6.2).

Table 6.2. Major engine tests, flight tests, and initial Constellation Program missions.

Test flight[1]	*Location*	*Year*	*Estimated number of tests/flights*
First stage ground tests			
Development Motor-1. Hot fire test		2009	1
Qualification Motor. Hot fire test		2011	2
Qualification Motor. Hot fire test		2012	1
Launch abort system tests			
Launch abort flight test		2009	1
Pad abort test		2010	1
Launch abort flight test		2010	1
Launch abort flight test		2011	2
Upper stage engine (J-2X) ground tests			
Upper stage engine hot fire test	Stennis Space	2010–2014	175
Upper stage engine hot fire test (simulated altitude)	Center	2010–2014	100
Upper stage engine hot fire test	Glenn Research Center	2011	2
Main propulsion test article hot fire test	Marshall Space Flight Center	2010–2013	24
Ares I *flights*			
Ares I ascent development flight test	Kennedy Space	2009	2
Ares I ascent development flight test	Center	2012	1
Orbital flight test		2013	2
Orbital flight test[2]		2014	2
Mission flight[3]		2015–2020	Up to 30
Ares V *core stage engine ground tests*			
RS-68B engine hot fire test	Stennis Space	2012–2018	160
Main propulsion test article cluster hot fire test	Center		20

Table 6.2 *cont.*

Earth departure stage engine ground test			
Upper stage engine hot fire test (simulated altitude)	Glenn Research Center	2012–2014	20
Main propulsion test article hot fire test	Marshall Space Flight Center	2015–2018	20
Ares V *flights*			
Flight test	Kennedy Space Center	2018	2
Mission flight[4]		2019	2
Mission flight		2020	1

1 The number, location, and types of tests are subject to change as Constellation test programs evolve.
2 The third orbital flight test will be the first crewed launch of *Orion/Ares I.*
3 Up to five *Ares I* flights per year will occur.
4 The second flight in 2019 is the first planned to include landing a crew on the Moon.

Within the structure of NASA, the Constellation Program is designed to evolve as near-term technical and programmatic objectives are realized, while remaining attentive to the lessons learned from current and prior programs. An example of this is the creation of the Shuttle and Apollo Generation Expert Services (SAGES), a pathway enabling the enlistment of retired experts from NASA's past, such as legendary Flight Directors, Chris Kraft and Glynn Lunney.

Rationale for returning to the Moon

> "A large portion of the scientific community in the U.S also prefers Mars over the Moon. But interest in the Moon is driven by goals in addition to and beyond the requirements of the science community. It is driven by the imperatives that ensue from a commitment to become a space-faring society, not primarily by scientific objectives, though such objectives do indeed constitute a part of the overall rationale."
>
> NASA Administrator, Dr Michael Griffin

Although the VSE's power is politically driven, the reasons for returning to the Moon, often referred to by mission planners as *drivers*, include science, technology, exploration, and exploitation.

Science

The potential of the Moon as a scientific outpost can be divided into the categories of science *of*, science *from*, and science *on* the Moon. Science *of* the Moon includes the potential for conducting geophysical, geochemical, and geological research, and enabling a better understanding of the origin and evolution of the Moon. Science *from* the Moon includes research that can be performed as a result of the unique properties of the lunar surface, such as astronomical observations on the far side of the Moon. Science *on* the Moon includes the disciplines of biology and exobiology that investigate the stability of biological and organic systems in hostile environments, and the regulation of autonomous ecosystems.

Technology

The hostile environment of the Moon, such as the radiation field, reduced gravity, and the ever prevalent dust, are very similar to the conditions on Mars and therefore offer a suitable test-bed to apply and evaluate technologies designed to deal with such an environment. Also of particular importance for future Mars missions is the testing of in-situ resource utilization (ISRU) on the lunar surface, since this process will be used by astronauts during Martian missions.

Other scientific objectives that must be met before undertaking manned missions to Mars include the development of autonomous tools and integrated advanced sensing systems, such as bio-diagnostics, telemedicine, and environmental monitoring. Once again, the lunar surface offers a demanding testing ground for the rigorous evaluation and development of such systems.

Exploration and exploitation

More than 2,000 people have set foot on Mount Everest, hundreds of adventurers have reached both the North and South Poles, and more than 500 spacefarers have experienced microgravity. New frontiers are opened up for the purpose of understanding the unexplored, and the next logical step in exploration is to venture beyond LEO and see the sights on our nearest neighbors. Given the resources of the Moon, it will not be long before industries develop infrastructures to extract life-support consumables, propellant, and Helium-3.

US motivations for realizing the vision

The Columbia effect

While each of the drivers mentioned represents a strong case for returning to the Moon, an equally strong justification can be made on the grounds that space is part of American culture, as evidenced by the national mourning that followed *Columbia*'s tragic disintegration in February, 2003. The US's reaction to the Columbia Accident Investigation Board's (CAIB's) conclusion that the accident was ultimately caused by a "failure of national leadership"[4] was President Bush's VSE. The VSE is an ambitious program intended to reverse a period of drought in which dreams had been limited and little investment had been made in building public or political support for space initiatives:

> "The human imperative to explore and settle new lands will be satisfied, by others if not by us. Humans will explore the Moon, Mars, and beyond. It's simply a matter of which humans, when, what values they will hold, and what languages they will speak, what cultures they will spread. What the United States gains from a robust program of human space exploration is the opportunity to carry the principles and values of western philosophy and culture along on the absolutely inevitable outward migration of humanity into the solar system and, eventually, beyond. These benefits are tangible and consequential. It matters what the United States chooses to do, or not to do, in space."
>
> NASA Administrator, Michael Griffin, AIAA speech

The Chinese motive

NASA's $200 billion plan for reinventing Apollo is also fuelled by the concerns echoed in the comments made by President Johnson more than four decades ago and the fear that China may beat the US back to the Moon. It took the US years to recover from the shock of being beaten into space by the Soviet Union, although America eventually prevailed, clinching the grand prize in the first space race by landing a man on the Moon. Now, some suggest the US will not be as lucky this time around. In the face of all the obstacles facing NASA, how much longer can the agency stay in the driver's seat?

Financial obstacles

The true cost of Orion and Ares

With more than $7 billion having been awarded to the Constellation Program, some observers are wondering how much it will really cost to develop the Shuttle's successor. However, even a preliminary cost estimate isn't possible until NASA has conducted each project's definition review, after which requirements may change, thereby affecting the cost. The Constellation Program's budget request maintains a confidence level of 65%, meaning that NASA is 65% certain that the actual cost of the program will either meet or be less than the estimate. However, given the historical cost overruns of past NASA systems such as the Space Shuttle and the known level of uncertainty of *Orion*'s requirements, it is likely cost estimates will increase over time.

Reducing costs

In an effort to reduce costs and minimize risk developing in the *Orion* and *Ares I/V* projects, NASA planned to maximize the use of heritage systems and technology, but, since 2005, changes have been made to the basic architecture designs for *Orion* and *Ares I* that have resulted in *decreased* use of these systems. Part of the reason for these changes is the ability to achieve even greater cost savings using other technologies and the inability to recreate heritage technology such as the Space Shuttle Main Engines (SSMEs). Other changes forced upon NASA as a result of weight issues include the increase in the number of segments on the *Ares I* SRB from four to five. While these changes partly contributed to the program's cost increases, analysts point to more worrying factors, such as schedule pressures, insufficient test facilities, and technology gaps.

Ares I schedule pressures

Once the Space Shuttle is retired in 2010, the US may face a five-year gap in manned spaceflight capability, during which time NASA astronauts will hitch rides to the ISS onboard the aging Russian Soyuz, although one potential solution to avoiding this gap may be to utilize the services of private sector company SpaceX (Figure 6.7). For the Americans, such a gap is uncomfortable, since it causes considerable schedule pressure for both the *Ares I* and *Orion* projects, as NASA tries desperately to reduce the time between the Shuttle's retirement and *Orion*'s first flight. For example, the development schedule for the J-2X rocket engine is very aggressive, with just seven years scheduled for development start to first flight in 2012. By comparison, the development of the SSME took *nine* years. Obviously, since the J-2X is designed to power the *Ares I* upper stage, its development represents a critical path and if any delays in DDT&E occur in the J-2X schedule, there will undoubtedly be a ripple effect of cost and schedule impacts.

Orion/Ares I schedule pressures

The *Orion* project is already suffering from schedule pressures caused by a number of technical issues ranging from weight issues to the troubling matter of thrust oscillation in the *Ares I*. Thrust oscillation is caused by accelerating gas vortices from the rocket matching the natural vibrating frequencies of the rocket motor's combustion chamber – a combination that causes shaking. Thrust oscillation is a phenomenon occurring in all solid rocket motors (SRMs), and was observed during the first Space Shuttle launches, before being resolved. However, attempts to resolve the thrust oscillation problem on *Ares I*, such as the installation of mass dampeners, have caused an increase in weight and a reduction in safety as a result of having to remove other hardware used for system redundancy.

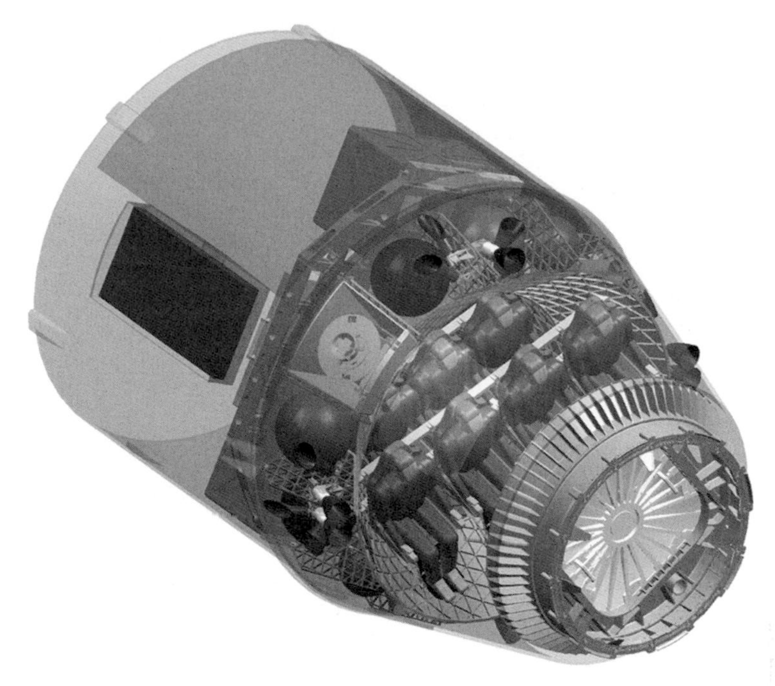

Figure 6.7 Artist's rendering of SpaceX's Dragon capsule in manned configuration. Image courtesy: SpaceX.

Test facilities

Another problem that will almost certainly adversely affect the *Ares I* and *Orion* schedules is the lack of appropriate testing facilities. For example, existing altitude test facilities are insufficient to test the J-2X engine, forcing NASA to construct a new altitude test facility at Stennis Space Center (SSC). An additional issue affecting the J-2X's development is the lack of test stands, since existing test stands are supporting the Shuttle program until 2010, although NASA is working to make an additional test stand available to J-2X contractors, Pratt and Whitney Rocketdyne.

The situation is not much better for *Orion*. One of the key test issues for the CEV is testing the thermal protection system (TPS). Unfortunately, the large-scale facilities for testing heat shields used during the Apollo era no longer exist, meaning that NASA has no ablative testing capability.

Technology and hardware gaps

Exacerbating the already tight development schedule are the technology and knowledge gaps associated with the hardware and technology that must be used in *Ares I* and *Orion* (Table 6.3). NASA also faces unwieldy difficulties such as the

Table 6.3. *Ares I* and *Orion* technology and hardware gaps.

	Ares I
First stage	• Thrust oscillation within the first stage causes excessive structural vibrations exceeding crew tolerances. A NASA "Tiger Team" is studying the issue and has proposed options, including vibration absorbers and redesigning elements of the *Orion* vehicle to isolate the crew from vibration • It is unknown how adding a fifth segment to the SRBs will affect flight characteristics. Incomplete understanding of flight characteristics may lead to risk of hardware failure and loss of vehicle control. Qualification requirements may be difficult to meet due to new ascent loads and vibration environments specific to *Ares I*
Upper stage	• One of the lessons learned from the Apollo program was not to use common bulkheads due to complexity and difficulties in manufacturing. Despite this, NASA redesigned the upper stage configuration from two separate propellant tanks to two tanks with one common bulkhead
J-2X upper stage engine	• The J-2X development will require 29 rework cycles in addition to extensive redesign to achieve increased performance requirements, and to meet human rating standards • Development of the J2-X faces significant schedule risks due to problems developing and manufacturing a carbon composite nozzle extension needed to meet thrust requirements
	Orion
Launch abort system	• Due to ongoing requirement changes related to *Orion* system and subsystems, development setbacks may prevent certain test objectives from being adequately demonstrated • Development setback is possible due to probable redesign of *Orion* for the launch abort system to address weight issues caused by thrust oscillation
Thermal protection system	• *Orion* requires the development of a large-scale ablative TPS, but, due to the size of the vehicle and the limited development time, a functional TPS may not be ready in time for *Orion* to be operational in time for ISS flights • A TPS of the size required by *Orion* has never been proven and must be developed. This requires time to develop technologies to maturity and meet TPS standards

weight and mass growth of both *Ares I* and *Orion*, which has already forced the agency to re-baseline the *Orion* design.

NASA's challenging transition

NASA's new course, as defined by the VSE, or the Obama version of it, has resulted in an agency engaged in a difficult and demanding transition. It is a transition requiring adjustments as NASA prepares to retire the Shuttle, sustain missions to the ISS, and prepare for returning humans to the Moon. Since NASA hasn't designed a new system for more than three decades, it is inevitable that relearning will be required. Unfortunately, many of those criticizing the program have very little understanding of the product lifecycles associated with developing new launch systems, most of which are longer than a two-year Congress, a four-year presidential, or a six-year Senate term. However, VSE was never designed to get products out of the door within politically significant timeframes. But, with China pursuing its own lunar exploration program and a new space race on the horizon, NASA may need to recalibrate to not only maintain the US's technological lead in space, but also to ensure there are no taikonauts on the lunar surface when American astronauts return!

CHINA'S MANNED SPACEFLIGHT PROGRAM

Manned spaceflight philosophy

Beijing's manned spaceflight ambitions have been articulated over the years in various government documents and policy papers, none of which has given much away. However, on November 22nd, 2000, China published a White Paper describing its future in space, although few details pertaining to launch schedules and exciting projects were revealed. Rather, the 13-page White Paper presented a series of bland statements intermingled with bureaucratic language designed more to exasperate than to inform. However, by reading between the lines and searching for nuances, it was possible for analysts to gauge the long-term priorities for China's manned spaceflight program.

Following China's third successful manned spaceflight in September, 2008, it is obvious the country is on a steep learning curve, closing one technology gap after another. The achievements of Chinese taikonauts are inevitably accompanied by pronouncements of space laboratories, space stations, lunar bases, and even manned Mars missions. These statements naturally make one wonder what Beijing is up to, but attempting to forecast which direction China's space program may take is compounded by the cloak of secrecy shrouding the Chinese system. Unlike the details of NASA's VSE, which can be found on thousands of websites, the goals of the Chinese National Space Administration (CNSA) are mired in ambiguity, contradiction, and confusion. Although the Chinese are adamant they will build a sustained manned spaceflight program, and not just plant a flag on the Moon (alluding to NASA's abandonment of its manned lunar program in 1972), the pathologically reticent Chinese are loathe to assigning dates to ambitions.

While one motivation is surely to beat the US to the Moon, another lesser objective is the mining of Helium-3.* With energy demand soaring into the

* Helium-3 is a light, non-radioactive isotope of helium that may be processed into a fuel for commercial fusion. Helium-3 is thought to have been deposited in the upper layer of regolith of the Moon by the solar wind over billions of years.

foreseeable future, China's energy experts are especially keen to find an alternative source. One possibility may be Helium-3, an isotope of helium that is especially abundant on the Moon.

Review of China's manned space program

China's program to launch humans into space has been active for more than 15 years and comprises three phases that follow Five-Year Plans. Phase I, which encompassed the 2001–2005 Five-Year Plan, included the first five Shenzhou flights, culminating in China's first manned spaceflight in October, 2003. To date, Phase II has included Shenzhou 6, which flew two taikonauts on a five-day mission in October, 2005, and Shenzhou 7, which flew three taikonauts and featured the country's first extravehicular activity (EVA). The remaining objectives of Phase II, which may be achieved within the 2006–2010 Five-Year Plan, include establishing a space laboratory module with docking capability and conducting a R&D task. Phase III objectives are less defined, but include establishing a permanent space station – a goal that may be achieved in the 2011–2015 Five-Year Plan.

Future Five-Year Plans

During 2007, the CNSA began to discuss conceptual manned space programs as a prelude to the next Five-Year Plan. Among the programs discussed was the Chinese Manned Lunar Program and lunar base after 2020 – a date that would coincide with one of the VSE's goals. Also examined was the possibility of establishing a 20-tonne manned space station. Both of these programs would be dependent on successful completion of the Long March (LM)-5 launch vehicle, but even when the LM-5 is declared operational, it is unsure whether either of these programs will be realized, since neither of them is officially a part of any existing state plan! However, precedence seems to have been given to the space station concept, since mock-ups have already been completed, whereas the idea of a manned lunar mission seems to have not taken hold as a political priority yet.

Manned lunar mission

One person who believes a Chinese manned lunar mission is firmly on the political agenda is former NASA Administrator, Dr Michael Griffin. At a Washington luncheon speech on the Space Economy on September 17th, 2007, Griffin explained to a surprised audience that he thought taikonauts would be back on the Moon before American astronauts. Griffin clarified his statement later by explaining that a manned lunar mission could easily be achieved without developing a Saturn V-class launch vehicle like NASA's planned *Ares V*. In fact, once China develops the LM-5 launch vehicle in 2014, it could quickly conduct a manned lunar circumnavigation mission followed shortly thereafter by a manned lunar orbit mission. These missions would then set the stage for a mission to land taikonauts on the surface of the Moon before 2020 – a date within the 2016–2020 Five-Year Plan.

Lunar precursor missions

The Shenzhou 7 mission of September, 2008, which showcased China's first EVA, almost certainly served as a prelude to more advanced Earth orbital R&D operations that will ultimately lead to landing taikonauts on the Moon. The next technological objective will feature the docking of two spacecraft. This goal will probably be realized during the unmanned Shenzhou 8 mission, scheduled for launch by the end of 2010, which will R&D with the *Tiangong 1* space laboratory module to create a rudimentary space laboratory, several of which may be launched during the 2011–2015 Five-Year Plan. The *Tiangong 1* space station will probably be crewed by taikonauts launched on Shenzhou 9 and Shenzhou 10. The Tiangong modules are approximately 8 tonnes and are believed to be lengthened orbital modules featuring front and possibly a rear docking port. Some believe the Tiangong module is a variant of the Soviet-era TKS module but, given the much smaller size, a more accurate comparison would be to describe it as a scaled-down version of the Salyut class space station concept.

Tiangong mission sequence

Establishing the *Tiangong 1* space station would commence with the unmanned launch of the *Tiangong 1* space laboratory module with a Shenzhou instrument propulsion module. This launch will most likely be followed by an unmanned and automatic docking demonstration of the Shenzhou 8 spacecraft. It is possible the Shenzhou 8 orbital module may serve as an instrument propulsion module for the *Tiangong 1*, in which case the *Tiangong 1* will jettison its own instrument propulsion module. The Shenzhou 8 spacecraft will also bring supplies for the two subsequent Shenzhou manned missions.

The Shenzhou 9 and Shenzhou 10 missions, each carrying two taikonauts, will probably occur within a month of one another, with the Shenzhou 10 mission being launched shortly before the end of the Shenzhou 9 mission. Before the Shenzhou 10 mission is launched, the Shenzhou 8 orbital module will be filled with trash, undocked from the *Tiangong 1* station, and de-orbited, thereby opening a docking port for the Shenzhou 10 spacecraft.

Shortly after the Shenzhou 10 spacecraft docks with the *Tiangong 1*, the Shenzhou 9 descent module, with its instrument propulsion module, will depart, leaving the Shenzhou 9 orbital module to serve as an instrument propulsion module. At the end of the Shenzhou 10 mission, the crew will return using their instrument propulsion and descent module, leaving the Shenzhou 10 orbital module to control the *Tiangong 1* station.

With the successful completion of the Tiangong/Shenzhou 8–10 missions, China will have demonstrated R&D, re-supply, extended duration, and interchangeable hand-off of the service control modules in addition to utilizing a manned space laboratory for multiple operations.

Mission timelines

The unmanned Tiangong and Shenzhou 8 missions are planned to launch before the end of the current 2006–2010 Five-Year Plan, while the Shenzhou 9 and Shenzhou 10 missions will almost certainly take place early in the next 2011–2015 Five-Year Plan. Assuming these missions are successful, China will be in a strong position to proceed to Phase Three, a step contingent on the development of the LM-5.

Long March-5

Although the LM-5 program was not officially launched until 2007, research and development of relevant technologies had already begun in 2000. The 120-tonne-thrust liquid oxygen (LOX)/kerosene core engine and the 50-tonne-thrust LOX/ liquid hydrogen (LH2) engine, which represent the key technologies of the launch vehicle, were successfully tested in 2005. The theoretical evaluation of the launch vehicle was completed by 2006 and the program definition phase was completed in 2008, with production of the first rocket beginning shortly thereafter.

The LM-5 will provide the Chinese with the necessary increased payload capacity to not only conduct manned lunar landing missions, but also to sustain future Chinese space program goals such as unmanned and manned missions to Mars. However, in the near term, operational status of the LM-5 is important if the goals of China's lunar program are to be realized. Currently, since China is driven by military requirements, priority is being given to Tiangong, but once the LM-5 is operational, the Chinese will increase their manned lunar effort.

China's lunar program

According to various reports, and depending on whom you believe, a Chinese manned lunar mission (Figure 6.8), including an EVA on the surface of the Moon, has been scheduled for 2024. In 2006, China's lunar program vice-director, Long Lehao, explained to reporters that the lunar EVA would be the culmination of a program that would include a lunar sample return mission and robotic precursor missions such as Chang'e, China's first lunar orbiter. Although Lehao's comments have been discounted by Sun Laiyan, chief of the CNSA, and Luan Enjie, chief of China's lunar orbiter project, China routinely denies a manned Moon landing plan. In reality, the success of Shenzhou has emboldened the Chinese to pursue the goal of landing taikonauts on the Moon. This objective was hinted at during Expo 2000 in

Hannover, where the centerpiece of the Chinese pavilion was a display of two taikonauts planting the flag of the People's Republic on the lunar surface. The Hannover exhibit was accompanied by declarations by Zhuang Fenggan, vice-chairman of the China Association of Sciences, that China would create a permanent lunar base with the intent of mining Helium-3. The following is a possible timetable China may follow to realize those goals.

Phase 1 would include lunar flyby or orbiting satellite missions, such as Chang'e, while Phase 2, characterized by unmanned soft-landing missions, would commence during the 2011–2015 Five-Year Plan. The soft-landing missions would test technology for robotic exploration using surface rovers – an objective that would be achieved during Phase 3 that might occur during either the 2016–2020 or the 2021–2025 plans. Finally, Phase 4 would include a lunar sample return mission, after which manned flight and construction of a lunar base would begin.

Once the LM-5 becomes operational, the Chinese will have all the hardware they need to embark upon a manned mission. The Shenzhou spacecraft re-entry capsule was modeled on the Russian Soyuz, which was designed and flight qualified in the 1960s specifically for return to the Earth from the Moon. In addition to the Shenzhou spacecraft, the Chinese would need a lunar lander and a LOX/LH2 stage to propel the configuration to the Moon. Launching such a payload into LEO would be within the capability of the LM-5. Two LM-5 launches would deliver the lunar injection stage and a Shenzhou-derived lunar lander to LEO. After docking with the booster stage, the Shenzhou would be launched on a trajectory to a direct landing on the Moon. This architecture was demonstrated by the Russians in the 1970s to be the most effective means of emplacing and supporting a lunar base (the lunar orbit R&D architecture employed by the US restricted the Apollo mission to locations near the lunar equator).

The technological capabilities required to successfully conduct a manned lunar mission may seem ambitious but, as we have seen, China's pursuit of a manned space program has pushed the country up the technical learning curves very quickly. With their first manned spaceflight, China achieved breakthroughs in 13 key technologies, including re-entry lift control of the manned spacecraft, emergency rescue, soft landing, malfunction diagnosis, module separation, and heat prevention. With their second and third missions, the Chinese mastered new skills such as EVA, and, with the Shenzhou 8, 9, and 10 missions, they will surely become proficient in the skills required to perform R&D procedures.

While China is progressing in leaps and bounds, like NASA, it is not immune to fiscal restraint and financial reality. China's manned space program is a stretched out multifaceted program that appears to be trying to accomplish too much with limited funds. While preparing Tiangong and follow-on Shenzhou missions, China is also trying to develop the LM-5, which is a huge drain from the funds available for the space program. Ultimately, there is only so much that can be accomplished in each Five-Year Plan and bankrolling a new launch vehicle, building a space station, *and* developing a manned lunar architecture have brought the Chinese space program to the brink of bankruptcy. While they plan to turn the manned Moon mission into a cost-effective enterprise, they will only make a profit if a way to create

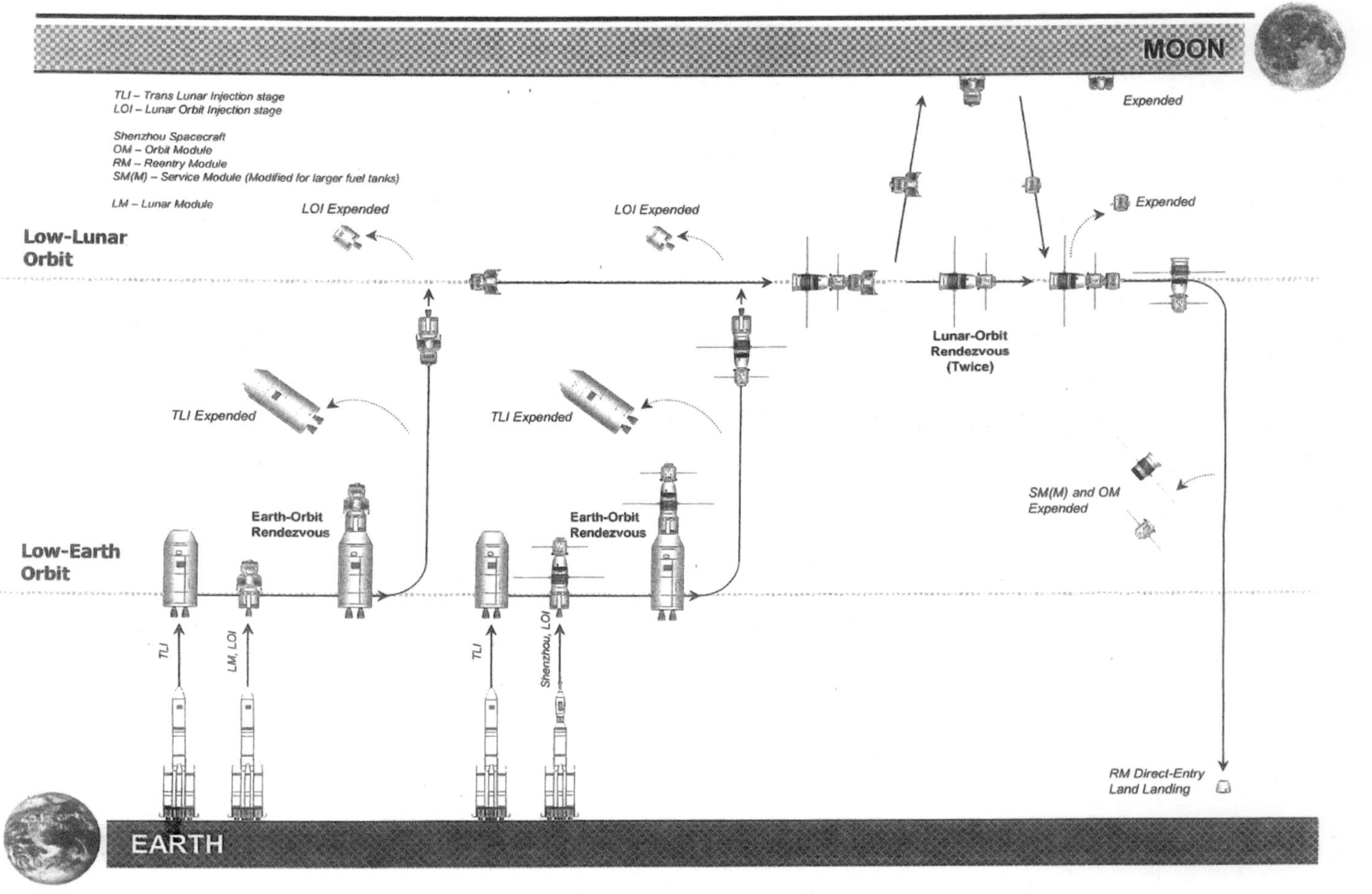

Figure 6.8 China's lunar mission architecture. Courtesy NASA.

fusion with Helium-3 is found. Furthermore, while China's space budget is continuously growing, their budget is still only a tenth of NASA's annual budget, and is woefully inadequate when one considers the number of technologies still required for China to mount a realistic challenge in the new race to the Moon.

National rivalries

The ultimate aim of Beijing's manned spaceflight program may not be as clear-cut as the VSE. However, it would seem the US and China are already in the early stages of a new space race that appears to have some of the hallmarks of the one between Washington and Moscow during the Cold War, when LEO was a proxy battleground for geopolitical dominance. While it may seem that the US has a clear advantage over the Chinese with its clearly defined plan for returning humans to the Moon, it remains to be seen whether the US can afford an independent program *and* maintain the requisite political will to fund it through to completion. Equally, at the risk of losing face to the Chinese, and allowing a technology gap to close, the US could push China towards increased spending on its manned program at a faster pace than it would otherwise choose. Such a policy might ultimately divert funds from Beijing's military programs and possibly cause China economic hardship similar to what happened when the US bankrupted the Soviet Union during the first space race. However, having a robust program is only one requirement for planning an incremental future for humans in space, whether they are astronauts or taikonauts. In a race as competitive as the one being fought in an environment as unforgiving as the cold hard vacuum of space, the outcome will likely be determined not just by a well planned program, but also by hardware and experience, which form the subject of the next two chapters.

REFERENCES

1. Treaty on Principles Governing the Activities of States in the Exploration and Use of Outer Space, including the Moon and Other Celestial Bodies (October 10, 1967).
2. www.nasa.gov/pdf/55583main_vision_space_exploration2.pdf.
3. NASA Authorization Act of 2005, HR 3250 (December 30, 2005).
4. Columbia Accident Investigation Board. NASA Government Printing Office, Washington DC (August, 2003).

7

Current and future hardware

When Moscow launched Yuri Gagarin into space in 1961, Soviet officials raised a veil of secrecy, awarding medals in private to those responsible for the feat. The Soviets also lied about the launch location and even deceived the Western press by saying Gagarin had landed in his capsule when, in reality, he had ejected from his spacecraft at an altitude of almost 4 km, before parachuting back to Earth. While a high level of secrecy also shrouds the Chinese space program, the degree to which they conceal the details of their space efforts seems to be less obsessive than the Soviets, as evidenced by websites promoting their space hardware and the brochures publicizing their rockets that government agencies hand out at conferences. Thanks to this information dissemination, Western experts have learned a great deal about Chinese space technology and the nuances of each spaceflight. In addition to revealing details of their spacecraft, information released by the Chinese shows that they have worked hard to make their launch vehicles reliable and that their manned space effort is no "man-in-the-can" program, as was the case in the American Mercury Program. With their rapidly improving hardware, the Chinese expect to chart a bold course that they hope will see them beat the US to the Moon. Meanwhile, on the other side of the Pacific, the US is working hard on realizing the goals set by the Vision for Space Exploration (VSE).

NASA'S HARDWARE

The path outlined by the VSE requires NASA to transition from an agency focused on low Earth orbit (LEO) operations to developing transportation modes capable of exploring the Moon, Mars, and beyond. As this book is being written, NASA is operating under a Bush Administration (continued under the Obama Administration) directive to complete assembly of the International Space Station (ISS) and retire the Space Shuttle fleet by 2010. After 2010, the US may need to rely on Russia to launch American astronauts (at a cost of $47 million per astronaut!) until *Ares I*, *Ares V*, and *Orion* are operational. However, the US is hoping the hiatus between the Shuttle's retirement and the operational status of *Orion* can be reduced by hitching a ride with SpaceX onboard their *Dragon* capsule.

During his election campaign, Obama indicated his interest in reviewing the Space Shuttle retirement plan once he took office. However, according to former NASA Administrator, Dr Michael Griffin, operating the Space Shuttle fleet beyond 2010 would probably *widen* the five-year gap, since *Orion*'s first flight in 2015 is based on diverting money saved from the Shuttle retirement to the development of the new vehicles.

Constellation hardware

Ares I and Ares V

Ares I (Figure 7.1) is the launch vehicle that will carry *Orion* into orbit, while the unmanned Cargo Launch Vehicle (CaLV), christened *Ares V* (Figure 7.2), will carry an Earth Departure Stage (EDS) and the Lunar Surface Access Module (LSAM), called *Altair*, to the Moon. *Ares*, the ancient Greek name for the Red Planet, was chosen to reflect the choice of Mars as one of the VSE's intended destinations, while the *I* and *V* designations pay homage to Apollo's *Saturn I* and *Saturn V* that enabled NASA astronauts to land on the Moon.

Ares I

Ares I is an in-line, two-stage rocket. The *Ares I* first stage is a five-segment RSRB that burns a specially formulated solid propellant called polybutadiene acrylonitrite (PBAN). Above the first stage (Figure 7.1) sits a forward adapter/interstage designed to interface with the *Ares I* liquid-fuelled upper stage. Above the interstage is a forward skirt extension housing the Main Parachute Support System (MPSS) and main parachutes for recovery of the first stage. The frustrum, located at the top of the first stage, provides the physical transition from the smaller diameter of the first stage and the larger diameter of the upper stage. The upper stage is based on the internal structure of the Space Shuttle's External Tank (ET). The upper section of the upper stage includes a spacecraft adapter (SCA) system designed to mate with *Orion*, while the lower section includes a thruster system to provide roll control for the first and upper stages. The Instrument Unit Avionics (IUA) provides guidance, navigation, and control for the entire launch vehicle. The IUA includes subsystems, such as the J-2X engine interface, upper stage reaction control system (RCS), Hydraulic Power Unit Controller (HPUC), Data Acquisition Unit, and the Ignition/Separation Unit. Power for the subsystems is provided by the Electrical Power System (EPS), comprising batteries, power distribution and control units, DC-to-AC Inverter Units, and cabling. Located in the IUA, the EPS ensures redundant sources of 28-volt direct current (VDC) from the time ground power is removed prior to launch until the end of the mission.

To prevent combustible LOX and LH2 accumulating to dangerous levels while *Ares I* is on the launch pad, the upper stage is fitted with a Purge System. This system

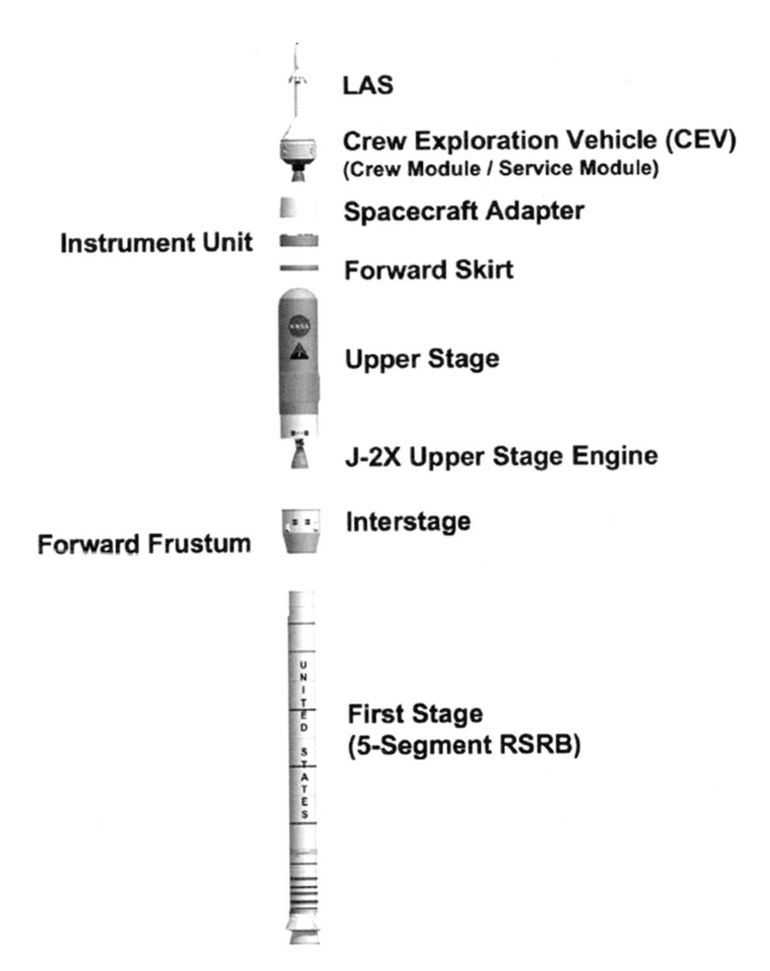

Figure 7.1 Exploded view of *Ares I* launch vehicle. Courtesy NASA.

ensures thermally conditioned inert nitrogen gas is pumped into the closed compartments of the upper stage while simultaneously exhausting excess nitrogen through special vents at the bottom of the compartments.

Ares I nominal mission profile

When operational, a typical mission will commence with the *Ares I* first stage boosting *Orion* to an altitude of 50 km, whereupon the first stage will separate and fall back to the ocean to be recovered in the same manner as the Shuttle's SRBs. The upper stage will then carry *Orion* to an elliptical orbit of 245 km. Once the upper stage separates, *Orion*'s propulsion system will power the spacecraft to its 300-km circular orbit. There, the crew module (CM) and the service module (SM) will either

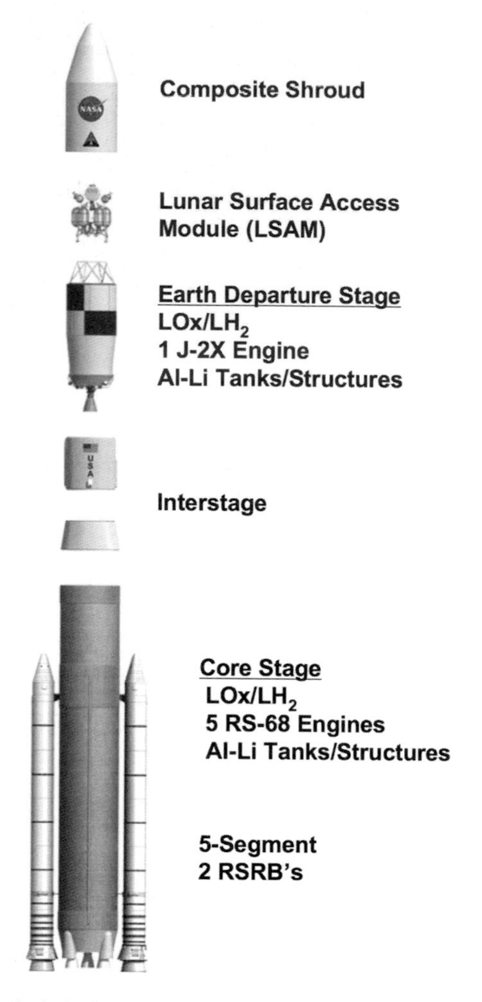

Figure 7.2 Exploded view of *Ares V* launch vehicle. Courtesy NASA.

rendezvous and dock (R&D) with the ISS or continue to the Moon, depending on the mission. Before this flight occurs, however, a series of test flights will be conducted.

Ares test flights

The first test flight (scheduled for October, 2009), designated *Ares I-X*, will utilize a flight test vehicle (FTV) similar in mass and configuration to the operational vehicle. *Ares I-X* will incorporate both flight and mock-up hardware specific to the objectives of the test flight, although it will not have the capability to receive ground commands while in flight. After lift-off from Kennedy Space Center (KSC), the FTV will climb

Figure 7.3 Launch abort system. Courtesy NASA.

to an altitude of 76,000 m. At 132 seconds into flight, first stage burnout and upper stage separation will occur, whereupon the upper stage simulator and the *Orion* crew vehicle and Launch Abort System (LAS) simulator will separate from the first stage (Figure 7.3). The simulator hardware will fall into the Atlantic and will not be retrieved, whereas the first stage booster will "fly" a recovery sequence, after which the hardware will be retrieved and inspected.

The second test flight, designated *Ares I-Y*, is scheduled for 2012. *Ares I-Y* will demonstrate flight control algorithms for the five-segment SRB and a high-fidelity upper stage simulator. It will also demonstrate a LAS at high altitude and measure and characterize launch and ascent environments for the five-segment SRB. If all goes well, an unmanned flight, designated *Orion 1*, will be launched in early 2013, eventually leading to the first human crew being launched in 2015.

Ares I and V propulsion

A new engine, known as the J-2X, will power the upper stages of both *Ares I* and *V*. The J-2X is an evolved version of two historic predecessors: the J-2 engine that propelled the *Saturn IB* and *Saturn V* rockets, and the J-2S, a simplified version of the J-2 developed and tested in the early 1970s. The J-2X will ignite 126 seconds after lift-off, following separation of the vehicle's first stage at an altitude of 60 km. The engine will burn for 465 seconds, lifting *Ares I* to an altitude of 128 km. Following engine cut-off, *Orion* will separate from the upper stage, whereupon *Orion*'s engine will ignite to provide the propulsive power required to place the capsule into LEO, where it will dock with either the ISS or the EDS.

Ares V

Ares V will serve as NASA's primary vehicle for delivering large-scale hardware to LEO, such as the lunar landing craft and materials for establishing a Moon base. The heavy-lifting *Ares V* is a two-stage, vertically stacked vehicle capable of carrying more than 180 metric tonnes into LEO and more than 70 metric tonnes to the Moon. For insertion to LEO, the *Ares V* first stage will utilize two 5.5-segment reusable solid rocket boosters (RSBs) derived from the Space Shuttle. The *Ares V* Core Stage, measuring 10 m in diameter and 64 m in length, will be the largest rocket stage ever constructed and will be almost as long as the combined length of the *Saturn V*'s first *and* second stages. Powering the *Ares V* Core Stage will be a cluster of five Pratt & Whitney Rocketdyne RS-68 rocket engines (Figure 7.4), each capable of supplying 700,000 pounds of thrust. The RS-68 engine, the most powerful LOX/LH2 engine in existence, will be modified by a series of upgrades to meet NASA's standards.

Ares V concept of operations

NASA's planned return to the Moon will commence with the launch of *Ares V* from KSC with an *Altair* lander and EDS under its payload shroud. After *Ares V* completes one orbit, *Ares I*, carrying *Orion* and its crew, will be launched and, after four days' orbital loitering in LEO, the two vehicles will R&D. At this point, the EDS will be restarted to propel the *Orion*/*Altair* configuration to the Moon.

Ares V elements

The *Ares V* upper stage, more commonly referred to as the EDS, will be powered by the same J-2X engine powering the *Ares I* upper stage. For *Ares V* missions, the first time the J-2X will be ignited will be 325 seconds after launch, at an altitude of 120 km, following separation of the *Ares V* first stage from the EDS and LSAM. Following a burn lasting 442 seconds, the EDS and LSAM will be placed in LEO. The second ignition of the *Ares V* J-2X will occur once *Orion* has docked with the EDS and LSAM. Once these flight elements are mated, the J-2X will fire for a second time, providing sufficient power to accelerate the mated vehicles to the escape velocity required for the *Orion*–EDS–*Altair* combination to break free of Earth's

Figure 7.4 RS-68 rocket engine. Courtesy NASA.

gravity and enter translunar injection (TLI). Once the mated vehicles arrive in lunar orbit, the *Orion*–lunar lander combination will jettison the EDS and its J-2X engine will perform a maneuver that will send these two flight elements into orbit around the Sun.

Figure 7.5 *Altair* attached to the *Orion* capsule. Courtesy NASA (*see colour section*).

Altair

Although the vehicle that will land astronauts on the lunar surface is still in the design process, the current vehicle iteration (Figure 7.5), designated the *711-A*, is probably close to the final design. *Altair* (Table 7.1) will be capable of transporting four crewmembers to and from the lunar surface, allow astronauts to return to Earth at any time during a mission, *and* be capable of landing 20 tonnes of cargo. In order to provide these capabilities, *Altair* will have variants for sortie, cargo, and crewed outpost missions.

Altair concept of operations

Mounted atop an EDS, *Altair* will be launched to LEO aboard *Ares V*. Once mated, *Altair* and the EDS will commence a loiter period of up to 95 days in LEO and wait for the crew in *Orion*, who will be launched atop *Ares I*. *Orion* will dock with *Altair* and the EDS will perform a TLI burn to place the *Altair–Orion* configuration on an Earth–Moon trajectory. Upon completion of the TLI burn, the EDS will be discarded. When the *Altair–Orion* configuration arrives at the Moon, *Altair* will perform a lunar orbit insertion (LOI) burn, and, following a loiter period in low lunar orbit (LLO), the crew will transfer aboard *Altair* and descend to the lunar surface.

Table 7.1. *Altair* characteristics.

Lander performance		*Vehicle concept characteristics*	
Crew size	4	*Ascent module*	
LEO loiter duration	14 days	Diameter	2.35 m
Launch shroud diameter	8.4 m	Mass @ TLI	6,128 kg
Lander design diameter	7.5 m	Main engine propellants	N_2O_4/MMH
Surface stay time	7 days sortie 180 days (outpost)	Number of main engines/type	1/derived OME/RS18 pressure-fed
Launch loads	5 g axial, 2 g lateral	Useable propellant	3,007 kg
Crewed lander mass (launch)	45,586 kg	Main engine Isp (100%)	320 sec
Crewed lander mass @ TLI	45,586 kg	Main engine thrust (100%)	5,500 lbf
Crew lander payload to surface	500 kg	RCS propellants	N_2O_4/MMH
Crew lander deck height	6.97 m	Number of RCS engines/type	16/100 lbf
Cargo lander mass (launch)	53,600 kg	RCS engine Isp (100%)	300 sec
Cargo lander mass @ TLI	N/A	*Airlock*	
Cargo lander payload to surface	14,631 kg	Pressurized volume	7.5 m^3
Cargo lander height	6.97 m	Diameter	1.75 m
Crew lander LOI *Delta V* capability	891 m/sec	Height	3.58 m
Cargo lander LOI *Delta V* capability	889 m/sec	Crew size	2+
Crew descent propulsion *Delta V* capability	2,030 m/sec	*Descent module (crewed)*	
Cargo descent propulsion *Delta V* capability	2,030 m/sec	Mass @ TLI	38,002 kg
TCM *Delta V* capability (RCS)	2 m/sec	Main engine propellant	LOX/LH2
Descent orbit insertion capability (RCS)	19.4 m/sec	Useable propellant	25,035 kg
Descent and landing reaction control capability	11 m/sec	Number of main engines Isp (100%)	1/RL-10 derived
Ascent *Delta V* capability	1,881 m/sec	Main engine Isp (100%)	448 sec
Ascent RCS *Delta V* capability	30 m/sec	Main engine thrust (100%)	18,650 lbf
		RCS propellants	N_2O_4/MMH
		Number of RCS engines/type	16/100 lbf each
		RCS engine Isp (100%)	300 sec

Orion

The role of transporting astronauts to and from the ISS, the Moon, and, ultimately, Mars will be met by *Orion*, a spacecraft with more than a passing resemblance to the Apollo capsule. In addition to providing habitable volume and life support for the crew, *Orion* features a docking capability and a means of transferring the crew to the LSAM. On return to Earth, a combination of parachutes and a water flotation system will be deployed for a water landing. Upon recovery, *Orion* will be refurbished and made ready for its next mission.

Orion (Figure 7.6) comprises four functional modules. On top of *Ares I* sits the Spacecraft Adapter (SA), which serves as the structural transition to the *Ares I* vehicle. Above the SA is the unpressurized service module (SM), providing propulsion and electrical power to the pressurized Crew Module (CM)/*Orion*, designed to transport either crew or cargo. Above *Orion* sits the Launch Abort System (LAS), which provides the crew with an emergency escape system during launch.

There will be four versions of *Orion*: Block 1A for transferring crew and cargo to the ISS, Block 1B, a pressurized manned version for transferring cargo to and from the ISS, Block 2 for lunar missions, and Block 3 for Mars missions. Each (Table 7.2) variant is designed to be used up to 10 times.

Table 7.2. *Orion*.

Diameter	5 m	Service module engine thrust	33,362 N
Pressurized volume	20 m^3	Lunar return payload	100 kg
Habitable volume	11 m^3	Propellant mass	9,350 kg
Dry mass	14,045 kg	Landing weight	7,337 kg

Orion systems and subsystems

From a design perspective, an advantage of choosing *Orion*'s blunt-body design is the familiar aerodynamic design. Thanks to the experience of Apollo, *Orion*'s development generated ascent, entry, and abort level loads that were familiar to engineers and resulted in less design time and reduced cost. However, the shape is about the only design aspect that *Orion* shares with its illustrious predecessor, as it provides a much larger habitable volume and incorporates the very latest in avionics and life support technology. *Orion*'s pressure vessel structure uses Aluminum 2024 honeycomb sandwich for the face sheets and Aluminum 5052 for the honeycomb core – a combination of materials enabling the vehicle to withstand the 14.7 psia internal cabin pressure required for ISS missions. For R&D purposes, *Orion* is fitted with five double-paned fused silica windows, two forward-facing, two side windows, and a fifth window located within the side ingress/egress hatch. *Orion*'s thermal protection system (TPS) will be the Avcoat ablator system, originally used for the

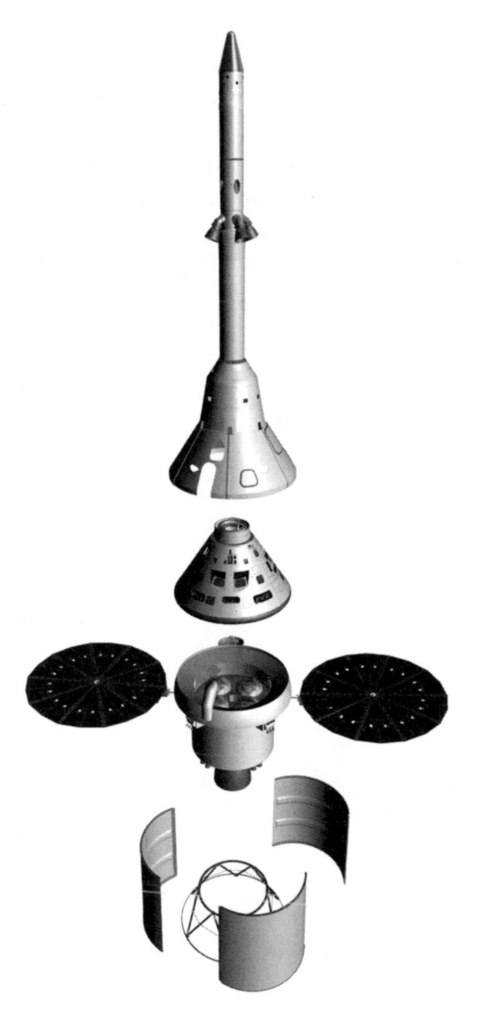

Figure 7.6 Exploded view of *Orion*. Courtesy NASA.

Apollo TPS and on select areas of the Space Shuttle during its earliest flights. Avcoat is made of silica fibers with an epoxy-novalic resin filled in a fiberglass-phenolic honeycomb and will be manufactured directly into *Orion*'s heatshield substructure.

Orion's propulsion system will comprise a Reaction Control System (RCS) that will include a number of elements such as the RCS tanks, the RCS pressurization system, the primary RCS thrusters, and back-up RCS thrusters. The RCS will enable *Orion* to perform exoatmospheric maneuvers and to orient itself during atmospheric re-entry. It will also provide astronauts with a means of counteracting induced spin and dampening induced pitch and yaw instabilities, which may occur during the lunar return trajectory.

The RCS propellant is a bipropellant system comprising Gaseous Oxygen (GOX) and liquid ethanol. The GOX mixture, which also feeds the life support system

Figure 7.7 *Orion* with its "Mickey Mouse" solar panels deployed, approaches the International Space Station. Courtesy NASA (*see colour section*).

(LSS), is stored in four cylindrical graphite-composite Inconel tanks mounted at the base of the vehicle, whereas the liquid ethanol is stored in two similar tanks and is pressurized by means of a high-pressure Gaseous Helium (GHe) system.

Three primary rechargeable Lithium-ion batteries, 28 VDC electrical power buses, power control units (PCUs), and back-up batteries will comprise *Orion*'s power subsystem, providing primary electrical power and distribution, and energy storage. The three primary batteries will be capable of providing 13.5 kW-hr for *Orion*'s two-and-a-quarter-hour storage requirement, the time between SM separation and landing. If more energy is required, there will be a fourth battery, providing one level of redundancy capable of providing 500 W of 28 VDC power for

45 minutes. The Lithium-ion batteries will feed electrical power to *Orion*'s power distribution system together with the two distinctive "Mickey Mouse" solar arrays mounted on the SM (Figure 7.7).

Vehicle communications

Command and Control (C&C) over all of *Orion*'s operations will be provided by the avionics subsystem comprising Command, Control, and Data Handling (CCDH), guidance and navigation, and communications. A part of the CCDH system will include two sets of translational/rotational/throttle hand controllers enabling the crew to take manual control of the vehicle when required. The equipment providing the crew with on-orbit vehicle attitude information, vehicle guidance, and navigation processing information will be supplied by the guidance and navigation system. At the heart of the system will be a Global Positioning System (GPS)/Inertial Navigation System (INS) that will work in conjunction with two star trackers, video guidance sensors, and two Three-Dimensional scanning Laser Detection and Ranging (LADAR) units.

Orion's avionics

Orion's flight control system will comprise three briefcase-sized Honeywell Flight Control Modules. In common with fighter aircraft cockpits, *Orion*'s control systems will rely heavily on *sensor fusion*, a type of automation relieving the astronaut-pilot of being a sensor integrator and allowing him/her to focus on the mission instead. Such a system makes sense, given that many astronaut-pilots are accustomed to advanced cockpits such as the F-15 and F-22. In *Orion*'s cockpit, pilots will be able to change displays as if they are revolving panels thanks to four flat-screen displays. During ascent to orbit, the displays will operate similarly to the screens in conventional airliners. One display will show an artificial horizon, another will display velocity, and a third will show altitude. The fourth display will show life support status and communications information. Once *Orion* reaches orbit, the displays will change to readouts showing R&D-related information such as the vehicle's flight path, range, and rate of closing.

Environmental control and life support system

Most of *Orion*'s Environmental Control and Life Support System (ECLSS) will be based on existing Shuttle technology or ISS systems and will include all the items necessary to sustain life and provide a habitable environment. Nitrogen gas required to sustain four crewmembers for nearly two weeks will be stored in cylindrical graphite composite Inconel 718-line tanks, while the oxygen will be stored in the four primary RCS oxygen tanks. Atmosphere regulation will be provided by a combined

Carbon Dioxide and Moisture Removal System (CMRS), ensuring that carbon dioxide levels are regulated, and an ambient temperature catalytic-oxidation (ATCO) system for contaminant control. Fire-detection and suppression capability will consist of spacecraft smoke detectors and a fixed halon fire-suppression system. Potable water will be stored in four spherical metal bellows tanks similar to the ones installed on the Shuttle.

Active thermal control system

Orion's ACTS will provide a temperature control capability for the vehicle consisting of a propylene glycol/water fluid loop with a radiator and fluid evaporator system. The fluid loop will work as a heat rejection system by using cold plates for collecting waste heat from the equipment, while a cabin heat exchanger regulates atmosphere temperature.

To deal with high heat loads, the ACTS will include a dual-fluid evaporator system that works by boiling expendable water or Freon R-134A in an evaporator. This cools the heat rejection loop fluid that is circulated through the walls of the evaporator, which, in turn, causes vapor to be generated and vented. The reason for a dual-fluid system is because water does not boil at the ATCS fluid loop temperatures and pressures at 30,000 m altitude or less, which means that from the ground to an altitude of 30,000 m, Freon R-134A is used.

Parachute and landing system

Orion will land in the Pacific Ocean near San Clemente Island, northwest of San Diego. However, in the event of an abort/contingency event, *Orion* will be capable of a land landing thanks to a wraparound partial airbag system comprising four cylindrical airbags located on the "toe" of the capsule. *Orion*'s automated parachute system will consist of two 11-m diameter drogue parachutes and three 34-m diameter round primary parachutes (Figure 7.8).

In the event of a land landing, *Orion* will be cushioned by four inflatable Kevlar airbags. As the vehicle descends, the airbags will be deployed out of the lower conical backshell. Two panels will jettison, permitting the airbags to inflate and wrap around the low-hanging corner of the heat shield to provide energy attenuation upon landing. Once *Orion* has landed, the airbags will vent at a specific pressure, facilitating a controlled collapse rate.

Service module

The SM's function will be to provide maneuvering capability, power generation, and heat rejection for *Orion*. The vehicle will feature a service propulsion system and a RCS comprising 24 thrusters, enabling it to conduct R&D with the LSAM in LEO.

Figure 7.8 *Orion*'s primary parachutes. Courtesy NASA.

Orion's power will be provided by two deployable, single-axis gimballing solar arrays using state-of-the-art junction photovoltaic (PV) cells. The reason for choosing solar arrays is the requirement for *Orion* to remain unoccupied in lunar orbit for up to 180 days – a period of dormancy considered too risky to rely on fuel cells. A Power Management and Distribution (PMAD) system will ensure that *Orion* receives adequate power, allowing for factors such as solar array degradation and losses incurred by the arrays not pointing at the Sun at the correct angle.

The SM will be a semi-monocoque unpressurized structure similar in design to the Apollo SM. The structure will provide attachment for *Orion*'s avionics, propulsion system components, and an interface for mating to the LM. A pressure-fed integrated-service propulsion system/RCS using LOX and Liquid Methane (LCH4) will comprise the SM propulsion. The propulsion system will be used for major translational maneuvers and vehicle attitude control, while minor translational maneuvers will be performed by 24 oxygen/methane pressure-fed RCS thrusters.

The SM will feature an avionics subsystem that will perform similar functions to *Orion*'s subsystem. The CCDH will include interface units that collect and transmit health and status data from other SM components, which will then be transmitted to *Orion*'s CCDH system. The SM's ACTS will comprise a single-loop propylene glycol fluid loop and a radiator mounted in *Orion*, except for the radiator panels fixed on the SM body.

Table 7.3. Spacecraft adapter.

Length	3.31 m
Basic diameter	5.03 m
Maximum diameter	5.50 m
Mass	581 kg

Spacecraft adapter

The SA (Table 7.3) will comprise a simple aluminum structure covered in a white silicone thermal control coating. The base of the structure will feature a field joint attaching to the first stage and vent holes that will equalize pressure during ascent.

Launch Abort System

The LAS is designed to pull *Orion* away from the thrusting *Ares I* first stage in the event of a contingency. The system incorporates an active tractor design utilizing a canard section below the attitude-control motor element. Below the canard section are four jettison motors sitting atop the systems interstage. Below the interstage is the abort motor element, comprising four exposed, reverse-flow nozzles. Attached to the aft end of the abort motor element is the adapter cone, which, in turn, is attached to the boost protective cover (BPC). Following second stage ignition, the LAS is discarded and abort contingency is provided by the SM propulsion system.

Orion abort modes

Orion's three abort modes are described in Table 7.4. Of the three abort modes, the most desirable is the ATO mode, since this allows the possibility of continuing a nominal mission or at least landing the crew within the continental US (CONUS), ensuring safer recovery operations.

Space suit systems

NASA's return to the Moon requires a versatile and rugged extravehicular activity (EVA) suit. Unsurprisingly, given the Moon-dust problems encountered by Apollo astronauts, the new class of astronauts will be hoping for a more robust and dust-resistant suit. To that end, engineers are developing the Constellation Space Suit System (CSSS). The Configuration One CSSS, designed for launch, landing, contingency events, and extravehicular activity (EVA), will resemble the current

Table 7.4. *Orion* abort modes.*

Abort mode	*Phase*	*Description*
Untargeted Abort Splashdown (UAS)	Abort initiation	Abort is initiated at *time zero* (t_0). *Orion*'s state is assigned based on interpolated CLV at t_0 and *Orion* coasts to re-entry interface
	Re-entry	CM separates from SM at re-entry interface and initial pitch angle is interpolated from trimmed aerodynamic database based on Mach number. The CM re-enters atmosphere and the bank angle is optimized so that abort initiation may occur as early as possible
Targeted Abort Landing (TAL)	Abort initiation	Abort is initiated at t_0 and *Orion*'s state is assigned based on interpolated CLV at t_0
	Separation	CM separates from SM and drifts for 15 sec
	Main engine burn	Main engine and auxiliary thrusters fired to boost downrange landing point into TAL recovery area, while maintaining altitude limitation of 121,200 m
	Re-entry interface	*Orion* coasts to re-entry interface 90,900 m
	Re-entry	*Orion* re-enters atmosphere, deploys parachutes at 15,150 m, and lands near TAL recovery area
Abort to Orbit (ATO)	Abort initiation	Abort is initiated at t_0 and *Orion*'s state is assigned based on interpolated CLV at t_0
	Separation	CM separates from SM and drifts for 15 sec
	Main engine burn	Main engine and auxiliary thrusters fired to boost *Orion*'s apogee altitude to 160 km
	Coast to apogee	*Orion* coasts to almost apogee
	Circularization	*Orion* circularizes using main engine

* Adapted from Falck, R.D.; Gefert, L.P. *Crew Exploration Vehicle Ascent Abort Trajectory Analysis and Optimization.* Glenn Research Center, NASA/TM-2007-214996.

Advanced Crew Escape Suit (ACES) used by astronauts on Space Shuttle flights. Unlike the current ACES suit, the Configuration One CSSS suit will feature a closed-loop environmental system (in common with a diving rebreather, the ACES suit uses an *open-loop* system, limiting operations above 30,000 m), new bearings in the shoulder, elbows, wrists, hip, and knees, and a full-pressure helmet with a swivel faceplate. The Configuration One suit will be made up of five layers, the innermost being a rubberized pressure-retaining bladder made of neoprene cloth, and the outer layer covered with Nomex in a fetching pumpkin orange.

The Configuration Two CSSS will be used primarily for lunar EVAs conducted during initial lunar sortie missions and later for the three to six-month outpost missions. This suit will comprise a completely new space suit, similar in appearance to the Mark III suit (Figure 7.9), designed and tested by ILC Dover in the early 2000s, but will incorporate the arms, legs, gloves, boot assemblies, and helmet from

Figure 7.9 Spacesuit engineer Dustin Gohmert simulates work in a mock crater of JSC's Lunar Yard while wearing the Mark III suit. Courtesy NASA.

the Configuration One variant. In common with the Mark III suit, the Configuration Two CSSS will incorporate a rear-entry hatch design, eliminating the need for the dual-plane closure used on the Shuttle/ISS EMU suits. Unlike the current NASA ACES suit, the Configuration Two CSSS will be a "soft-suit" design, enabling astronauts to bend over and grasp objects when fully suited and pressurized. New joint designs on the Configuration Two CSSS, together with the soft-suit elements from the Configuration One CSSS, will allow the suits to operate at a higher pressure (approximately 8 psia), thereby eliminating the danger of astronauts suffering from decompression sickness, normally associated with the lower pressure of the Shuttle/ISS suits, which operate at only 4.3 psia.

CHINESE HARDWARE

Long March launch vehicles

Long March-2F

The Long March (LM)-2F was developed in the 1990s to launch the Shenzhou manned vehicle. Between November, 1999, and September, 2008, the LM-2F successfully completed seven missions, before being phased out. Since the successors to the LM-2F, known as LM-2F/G and LM-2F/H, will be upgraded variants of LM-2F, the design and configuration of the LM-2F are discussed here.

Design

The LM-2F (Figure 7.10) comprises a two-stage core with four liquid-fuel strap-on boosters, a payload fairing, and an emergency launch escape system. The total length of the vehicle is 58.34 m. The core vehicle is powered by the YF6-2 engine, comprising four 75,500-kg-thrust YF5-1 chamber motors with swinging nozzles. The second stage of the core vehicle utilizes a YF20-1 rocket motor comprising one 75,600-kg-thrust main motor with fixed nozzles, a YF21-1 swiveling venire motor with four chamber motors, and propellant tanks.

The LM-2F incorporates 10 subsystems, including launcher structure, control system, power equipment, fault monitoring management system, escape system, remote monitoring system, safety external monitoring system, propellant utilization system, auxiliary system, and ground facilities. The overall length of the launch escape system (LES) is 15.1 m, with a maximum diameter of 3.8 m, and a weight of 11.26 tonnes. In the event of an emergency or major malfunction, the five solid rocket motors fitted to the LES would fire to pull the Shenzhou capsule and orbital module free from the launch vehicle and land in a safety zone using a parachute. The LES is available from 15 minutes prior to launch to the point of escape tower jettison 160 seconds (39,000 m altitude) following launch.

The launch vehicle is transported in several elements by rail to the Jiuquan Satellite Launch Centre (JSLC). At JSLC, the LM-2F is examined in the Horizontal Processing Building (HPB) before being moved to the Vertical Processing Building (VPB), where the core stage and strap-on boosters are lifted into the vertical position and finally assembled into a launch vehicle. Once the launch vehicle is assembled, it is integrated with the Shenzhou spacecraft – a process that normally takes between five and six weeks. Following additional testing and assessment, which usually take another two to three days, the complete spacecraft is moved on a mobile launch pad to the launch complex, and, following final checks performed on the launch pad, the vehicle is certified for launch (Table 7.5).

Long March 2F/H

The LM 2F/H will be the second-generation launch vehicle for the Shenzhou spacecraft. The LM 2F/H's first launch is expected to take place in the 2010–2012 timeframe, probably for the Shenzhou 9 mission, ferrying taikonauts to the space laboratory. In common with the LM 2F, the LM 2F/H will use a 3.35-m diameter core module and four 2.25-m diameter strap-on boosters, burning LOX and kerosene propellant. The core stage will consist of two 120-tonne-thrust YF-100 engines, while the strap-on boosters will consist of a single 120-tonne-thrust YF-

Figure 7.10 Long March-2F launch vehicle. Courtesy Federation of American Scientists.

Table 7.5. LM-2F flight sequence.

Time	*Event*
T + 0 sec	Launch vehicle lifts off
T + 10 sec	Launch vehicle pitches over
T + 2 min	Escape tower is jettisoned
T + 2 min 20 sec	Strap-on boosters jettison
T + 2 min 39 sec	First stage separates from launch vehicle
T + 3 min 20 sec	Spacecraft fairing is jettisoned
T + 7 min 41 sec	Second stage main engine shuts down
T + 9 min 40 sec	Second stage swiveling venire motor shuts down
T + 9 min 43 sec	Shenzhou spacecraft separates from the launch vehicle

100 engine. The LM2F/H will be available in two variants: a manned variant to launch the Shenzhou capsule (launch weight of 582 tonnes and a LEO payload capacity of 12,500 kg), and an unmanned variant to launch cargo spacecraft and space station modules (launch weight of 579 tonnes and a LEO payload capacity of 13,000 kg).

Long March-5

The LM-5 represents China's next-generation launch vehicle family, which will provide a range of launch vehicle configurations capable of supporting Beijing's ambitious manned spaceflight and lunar exploration plans. On October 30th, 2007, ground was broken for the LM-5 production plant in the Binhai New Area of Tianjin, a port city 120 km southeast of Beijing. The requirement for a new production facility was due to the large diameter of the LM-5's core stage, which would prevent it from being efficiently transported by rail or road to China's existing satellite launch centers in Xichang, Jiuquan, and Taiyuan. On completion, the facility will be capable of producing as many as 30 LM-5 rockets each year. The first launch of the LM-5 is likely to occur in Wenchang, in China's southern-most island province of Hainan, where a new satellite launch center is presently under construction and is scheduled to be operational in 2012.

Long March-5 design

The LM-5 family of vehicles will include three modular core stages. The overall length of the vehicle will be 60.5 m, with a launch weight of 643 tonnes, and thrust of 833.8 tonnes. Boosters with varying capabilities will be assembled from the modular core stages and strap-on stages. The modular design principle will be extended to engine selection, with various engines using different liquid propellants being utilized. Engine development of the liquid propellant YF-100 and YF-77 engines began in 2005 and, by 2007, versions of both engines had been successfully tested. Depending on the modular configuration of boosters and engines, the LM-5 will be capable of delivering up to 25 tonnes to LEO or up to 14 tonnes to geosynchronous transfer orbit (GTO).

Vehicle configuration

The LM-5 family will be based on three module types with diameters of 2.25, 3.35, and 5.0 m. The 2.25-m module will be powered by one 120-tonne thrust kerosene/liquid oxygen (LOX) engine, whereas two such engines would be used to boost the 3.35-m module. The 5-m core stage would use two 50-tonne-thrust liquid hydrogen (LH2)/LOX engines. The versatility of the modular approach means that any of the modules can be used as first stage units on small and medium launch vehicles. For example, the two smaller 2.25-m modules could serve a role as strap-on boosters for the larger 3.35 and 5.0-m core stage, in combinations of two or four.

Three upper stages, one in each of the three module diameters, are also planned. The 5-m upper stage would be powered by two LH2/LOX engines, each capable of producing 8 tonnes of thrust, whereas a new 15-tonne-thrust engine may be designed to power the 3.35-m upper stage.

Development of the LM-5 will probably focus initially on the 5-m variant, since existing LM vehicles already have the payload capability of the planned smaller 2.25 and 3.35-m versions. The 5-m diameter LM-5 will weigh 175 tonnes at lift-off and comprise a 31-m-long core filled with 158 tonnes of propellant together with four 2.25-m strap-on boosters. A more powerful version of the LM-5, designated CZ-5-504, may utilize four 3.35-m modules strapped onto the 5-m core – a configuration weighing as much as 800 tonnes and producing more than 1,000 tonnes of thrust!

Shenzhou spacecraft

China's most recent manned spaceflight was achieved using the Shenzhou spacecraft (Table 7.6), a dome-shaped vehicle comprising three modules. The basic design of the Shenzhou vehicle (Figure 7.11) is modeled after the Russian Soyuz vessel, first launched four decades ago. Capable of seating three taikonauts, Shenzhou's development began in 1992, under the auspices of Project 921-1, a designation of the Chinese National Manned Space Program. One of the objectives of Project 921-1 was to conduct a manned flight in October, 1999, prior to the new millennium. Although Shenzhou was launched in 1999, the first flight was unmanned. The first flight was followed by three additional unmanned test flights in 2001 and 2002, before manned launches on October 15th, 2003, October 12th, 2005, and September 25th, 2008.

Figure 7.11 Shenzhou-5 capsule close-up view. Courtesy Taikong Tansuo (*see colour section*).

Shenzhou modules

Shenzhou, in common with its Russian cousin, Soyuz, comprises three modules: a forward orbital module, a re-entry capsule in the middle, and an aft service module. The division of three modules is designed to minimize the amount of material returned to Earth. This design philosophy is identical to the one employed by the Russians, who put the crew and anything to be returned to Earth in a separate re-entry module. Everything else is put in the other two modules. By utilizing this method, the Russians minimize the need for heat shields, retropropulsion systems, parachutes, and other equipment required for landing. While the Shenzhou is based on the same principle as the Soyuz, it is not a copy, as evidenced by a much larger volume and an orbital module with its own propulsion system.

Table 7.6. Shenzhou spacecraft.

Shenzhou specifications			
First launch	November 19, 1999	Principal modules	Orbital module, re-entry vehicle, service module
First manned flight	October 15, 2003 (Shenzhou 5)	Overall length	9.25 m
Principal uses	LEO operations, space station ferry	Habitable volume	14.0 m^3
Performance	196 × 324 km orbit, 42.5°C	Launch mass	7,840 kg
Shenzhou module specifications			
Orbital module		*Re-entry module*	
Design life	200 days	Crew size	3
Length	2.8 m	Design life	20 days
Basic diameter	2.25 m	Length	2.50 m
Span	10.4 m	Basic diameter	2.52 m
Habitable volume	8.00 m^3	Habitable volume	6.00 m^3
Mass	1,500 kg	Mass	3,240 kg
Service module			
Design life	20 days	Mass	3,000 kg
Length	2.94 m	RCS propellants	N_2O_4/MMH
Basic diameter	2.50 m	Main engine	4 × 2,500 N
Span	17.00 m	Main engine thrust	10,000 kN

Orbital module

The orbital module contains space for experiments, crew-serviced equipment, and on-orbit habitation. Unlike the Soyuz, the Shenzhou orbital module is equipped with its own propulsion and control systems, enabling autonomous flight. Another difference between the Soyuz and Shenzhou designs is the option of leaving the Shenzhou orbital module in orbit for re-docking with another spacecraft – a capability that the Soyuz does not have because the hatch enabling it to function as an airlock is part of its descent module. Thanks to this design, the orbital module may be used in the construction of a Chinese space station.

Re-entry module

Shaped like an old-fashioned automobile headlight, the re-entry module is located in the middle section of the Shenzhou spacecraft. The capsule contains the spacecraft's instrument panel, limited storage space, and seats for three crewmembers. In common with the Command Module of the Apollo spacecraft, the Shenzhou re-entry capsule has no reusable capabilities.

Like the Soyuz cousin, the Shenzhou re-entry capsule employs the same landing technique, deploying a single drogue, followed by a single main chute. Another Soyuz trademark is the soft-landing system, which commences following jettisoning of the heat shield jettison, and is enabled by ignition of soft-landing rockets prior to impact.

The similarity between the Shenzhou re-entry capsule and the Soyuz spacecraft is no accident. In the mid-1990s, the Chinese purchased a complete Soyuz re-entry capsule from the Russian spacecraft manufacturer, RKK Energia. Although the Soyuz capsule sold by the Russians contained little hardware, it provided Chinese spacecraft designers with plenty of ideas. For example, in common with the Soyuz spacecraft, manual control of the Shenzhou re-entry capsule is performed using Soyuz-style hand controllers.

Constructed mostly of titanium, with aluminum alloy sidewalls, the re-entry vehicle features two small portholes, approximately 30 cm in diameter, providing a limited view outside. To orient the spacecraft manually for re-entry, an optical device similar to the one used on the Soyuz is used. The device features eight side ports for ensuring the spacecraft is correctly aligned for retrofire, but, unlike the Soyuz device, the Shenzhou version does not function as a periscope, which means that future docking activities will require control of the spacecraft from a separate station in the orbital module.

The capsule's habitable volume is limited to 6 m^3 by the housings for the main and back-up parachutes. The three crew seats face a high-technology control panel consisting of an array of flat panel screens, caution warning lights, and enunciator panels. The digital display system consists of two multi-functional color LCD displays, fed from redundant controllers. The LCD panels are high-resolution, capable of displaying complex graphics. From left to right, the crew seats are occupied by a payload specialist, a commander/pilot, and a co-pilot/flight engineer. Each crewmember is provided with a space suit derived from the Sokol suit used aboard Soyuz, a survival suit, underwear, an intravehicular (IVA) suit for working, a headset, medical monitoring sensors, a personal survival kit, an individually contoured seat liner, and a sleeping bag.

The ECLSS regulates the supply of nitrogen and oxygen to control the cabin pressure, circulates and purifies the cabin air, controls the cabin temperature and humidity control, and includes a fire monitoring and suppression system. The ECLSS maintains the cabin atmosphere at 81–101 kPa, oxygen partial pressure at 20–24 kPa, humidity at 30–70%, and temperature between 17 and 25°C, although, during re-entry, the cabin temperature can rise to 40°C. In the event of rapid/explosive decompression, a space suit loop connects the ECLSS to the crew using umbilicals connected to their space suits, thereby preventing asphyxia.

The guidance and navigation system determines spacecraft attitude and position using digital and analog sun sensors, infrared horizon sensors, an inertial measurement unit (IMU), and a global positioning system (GPS) receiver. All guidance and navigation system components are linked to triple-redundant majority-voting computers – a feature similar to the Space Shuttle.

Data transmission utilizes a Unified S-Band System, compliant to international standards, and combines telemetry, command, voice and video signals into one data stream. Location of the re-entry capsule is achieved by the use of three rescue beacons. Once the capsule reaches 40 km altitude, a 243-MHz VHF beacon starts transmitting while an HF transceiver transmits the capsule's precise location based on its GPS receiver. In the event of a landing remote from recovery forces, a 406-MHz international Emergency Position Indicating Radio Beacon (EPIRB) provides a method of locating the capsule.

The capsule contains 21 avionics black boxes, which communicate to one another as a distributed computer network using a dual-redundant military standard data bus. During flight, the Central Terminal Unit (CTU), a triple-redundant fault-tolerant computer located in the re-entry module, controls all three modules, but, following retrofire and separation of the other modules, it controls the re-entry vehicle during re-entry. As an additional level of redundancy, a second (dual-redundant computer) CTU located in the orbital module is designed to automatically come online if the primary system in the re-entry vehicle fails.

To control orientation following separation from the orbital and service modules and during re-entry, the re-entry vehicle is fitted with eight 150-N-thrust monopropellant hydrazine engines, pressure-fed from two rubber bladder tanks containing 28 kg of propellant. The sequence of events during re-entry is described in Table 7.7.

On the capsule's exterior are six antenna mounts, exhaust nozzles of the attitude control system (ACS) used during re-entry, and connections for gas, liquid, and electrical services.

Service module

The SM provides the electrical power, attitude, control, and propulsion for the spacecraft in orbit. As with the re-entry capsule, the SM bears more than a passing resemblance to the Soyuz, although it is longer than its Russian counterpart and its flared base is less pronounced. In common with the Soyuz spacecraft, the module's base is used as an external radiator surface for the spacecraft's thermal regulation system. Thermal regulation of the capsule is achieved by deploying two solar panels from the sides of the service module, but, unlike the panels utilized by Soyuz, the Shenzhou panels can be rotated to obtain maximum solar insulation, regardless of spacecraft attitude.

Roll and translation attitude of the spacecraft is achieved by the use of control engines positioned on the forward end of the SM. Within the base of the module, a series of thrusters firing from the outer rim of the flared base towards the central axis of the spacecraft control pitch and yaw.

Within the SM are several ellipsoid-shaped compartments housing avionics,

Table 7.7. Re-entry sequence of events.

Altitude (m)	*Time (hr:min:sec)*	*Event*
N/A	00:00:00	Orbital module separation
N/A	00:01:00	Main engine retrofire. The main engines on the Shenzhou's service module are ignited
N/A	00:04:00	Main engine shutdown
N/A	00:23:00	Service module separation. The service module separates from the re-entry capsule and is left to burn up in the atmosphere
120,000	00:00:00	Terminal area interface. Shenzhou's landing system becomes operational
10,000	00:36:00	Parachute jettison. Parachute hatch is jettisoned and two pilot parachutes are deployed. The pilot chutes pull out the drogue parachute, followed by the ring-sail main parachute. In the event of a parachute malfunction, a back-up parachute is used
6,000	00:39:00	Heat shield jettison. Heat shield is jettisoned, exposing a gamma ray altimeter and the four nozzles of the soft-landing retrorockets
2,000	00:44:00	Two-point suspension. Capsule converts from single-point to two-point suspension under the main parachute. Vertical velocity is reduced to 8 m/sec
1	00:48:00	Retrorocket ignition. Altimeter initiates ignition of four solid propellant retrorockets, reducing final impact velocity to 3.5 m/sec
Landing	00:48:01	Beacon initiation. Visual flashing beacon assists recovery forces in locating the capsule at night. If the capsule splashes down in water, fluorescent dye is released into the water to assist sighting from air

electrical, and ECLSS equipment. Also located in the SM are nitrogen and oxygen gases for the ECLSS, which are stored in steel alloy spheres. To reject the heat generated by Shenzhou's crew and electrical systems, a cooling system using heat pipes conducts heat from internal systems to an external radiator.

The SM's propulsion comprises a unified system feeding both attitude control and main engines from four 230-liter propellant tanks filled with 1,000 kg of N_2O_4/MMH propellants. The engines are pressure-fed using six 20-liter titanium cold gas tanks and a system of diaphragms located within the propellant tanks. The engines consist of four large expansion ratio main engines with a thrust of 2.5 kN each and a specific impulse of 290 seconds. Eight 150-N high-thrust pitch and yaw thrusters are located inside the flared section, eight low-thrust 5-N pitch and yaw thrusters located outside the flared section, and eight low-thrust 5-N roll/translation thrusters are positioned at the spacecraft's center of gravity.

Space suit systems

The Chinese have bought several space suits from the Russians. The suit worn by Yang Liwei during Shenzhou 5 resembles a Sokol-KV2 suit, although it is believed to be a Chinese-manufactured variant rather than an actual Russian suit. Following Liwei's historic flight, the Chinese signed a contract with Russia for three Orlan suits, two low-pressure training suits, and four suits for underwater training. In accordance with the contract, the Chinese are responsible for providing power and communications for some of the suits. During Shenzhou 7, one Chinese taikonaut wore the new EVA Orlan-M space suit and two wore the first-generation Feitian EVA, modeled on the Orlan-M.

Orlan space suit

The Orlan space suit (Figure 7.12) is a semi-rigid (Table 7.8) suit comprising a solid torso and helmet but flexible arms. It was designed with a rear hatch entry through the attached backpack, permitting quick donning (approximately five minutes). The exterior of the hatch houses the life support equipment, enabling up to nine hours' operation. The standard suit standard pressure is 0.4 atm, requiring a pre-breathe period of 30 minutes. Control of the suit is via a panel on the chest, with the markings in mirror image, enabling the cosmonaut to view the panel using a mirror on the suit's wrist.

While the Orlan-M and Feitian space suits share a common design philosophy, the Chinese EVA suit is equipped with a number of new features. For example, digital technology is applied extensively in the design of the Feitian. Another feature distinguishing the Feitian from the Orlan is the use of Code Division Multiple Access

Table 7.8. Technical characteristics of the Orlan-M space suit.

Nominal duration of the autonomous mode	7 hr $\sim$
CO_2 absorption cartridge operating time (airlock time included)	9 hr
Suit positive pressure	Nominal mode: 392 hPa Emergency mode: 270 hPa
Oxygen available (main and back-up)	1 kg
Cooling water	3.6 kg
Assured heat removal	Average: 350 W Maximum: up to 600 W
Total consumed power by the suit systems	Up to 54 W
Space suit weight (wet)	$\sim$109 kg
Service life	Up to 15 EVAs over 4 years

Figure 7.12 Orlan space suit. Courtesy NASA (*see colour section*).

(CDMA) technology – a data communication concept permitting several transmitters to send information simultaneously over a single communication channel. In comparison, the Orlan-M is constrained by short-wave communication.

An assessment of China's current space technology capabilities

China's LM family will remain the backbone of the space program for the near future, regardless of which path China chooses to follow. Equally, the Shenzhou spacecraft has proven to be a rugged and reliable vehicle capable of serving manned spaceflight goals for many years to come in much the same way as the Soyuz has served the Russians. While the ambitious goals to build a space station and land taikonauts on the Moon require plans for hardware that they do not yet possess, the Chinese plan has always been able to skip generations of technology simply by buying or absorbing foreign technologies. This policy has been employed very successfully through bilateral, multilateral, and more nefarious means, and is a policy that is unlikely to change in the future. Given their present state of technology and development of hardware, perhaps the only element that may prevent them attaining their goals is their lack of manned spaceflight expertise – a subject addressed in the following chapter.

8

Manned spaceflight experience

Sending humans into orbit no longer remains the niche capability possessed only by the US and Russia. China, having conducted three manned spaceflights and an extravehicular activity (EVA) since 2003, has entered the human spaceflight arena. In becoming a member of the astronaut club, Beijing has created additional political pressures on the US manned space program, which has suffered from declining budgets and downsizing during the last 20 years. Furthermore, achieving manned spaceflight capability has brought China additional prestige and geopolitical influence, which may be increased by competing in a race that the US won 40 years ago. With American policy seeking to isolate China on the possibility of cooperation, largely due to concerns about human rights and technology transfer, the stage seems set for another race to the Moon and a threat to American international leadership in human spaceflight.

There is no doubt that China has achieved impressive advances in human spaceflight with relatively few flights. For example, the US flew five orbital missions over 3.5 years before accomplishing its first EVA during Gemini 4 in June, 1965. The Soviet Union, by comparison, accomplished the first EVA on its seventh manned mission in March, 1965, with four years of human spaceflight experience. China took five years and did it in just three missions! It is a pace of development that seems likely to continue, as China now plans to establish a space station. On October 6th, 2008, during a panel discussion hosted by the American Association for the Advancement of Science (AAAS), China pronounced that they intended to fly three space station missions between 2010 and 2015. Furthermore, the space stations, known as Tiangong, would be visited by up to eight Shenzhou spacecraft. If accurate, this means that China plans to dramatically increase the pace of its human spacecraft launches. So far, China has averaged one manned launch every 2.5 years. By 2011, that rate could increase to one to two missions per year. The Chinese delegate attending the AAAS meeting explained that China would develop a manned space station by 2020, with the aim of establishing a long-term presence in space. It is all part of a three-stage strategy that aims to ultimately land taikonauts on the Moon. But can a country like China, a neophyte in the world of manned spaceflight, hope to compete with the US, which routinely chalks up more manned spaceflight experience in a week than the cumulative total of all China's missions combined?

CHINA'S MANNED SPACEFLIGHT EXPERTISE

Taikonauts and Yuhangyuans

When China successfully completed its first manned spaceflight in October, 2003, there was some confusion in the media about what to call the pilot, Colonel Yang Liwei. Since the dawn of the manned spaceflight era, the only two words we have had for space travelers are *cosmonaut*, from the old Soviet Union, and the more common term, *astronaut*, used to describe American and European spacefarers. A third descriptor originated in the late 1990s in reference to the Chinese manned space program. *Taikonaut* is the Chinese counterpart to astronaut and cosmonaut and is a term created from the Chinese word *Taikong*, meaning space or cosmos. The resulting prefix, *taiko*, is similar to *astro* and *cosmo*, whereas the *naut* ending derives from the Greek *nautes*, meaning sailor. While the term *taikonaut* seems to have been invented by space enthusiasts and adopted by journalists, a Chinese term has existed for several years to describe participants in the American and Russian space programs. The term *yuhangyuan* is a transliteration of Chinese words meaning literally *universe travel worker*. Since the term *astronaut* has been available for decades, it is unclear why so many English-language writers use the Chinese term, especially when the *China Daily* and the *South China Morning Post* both use *astronaut* in their English-language reports. More often than not, the use of the two terms is simply to distinguish between spacefarers of different origins. For this reason, *taikonaut* will be used when referring to Chinese space travelers and *astronaut* when referring to American space travelers.

Chronology of China's manned spaceflights

Shenzhou 5

Shenzhou 5 (Figure 8.1 and Table 8.1), the first manned spaceflight mission of the People's Republic of China (PRC), was launched on October 15th, 2003. The Shenzhou spacecraft was launched by a Long March (LM)-2F booster, following four unmanned test flights.

Shenzhou 5 was launched from Jiuquan Satellite Launch Center (JSLC), a launch site in the Gobi Desert. The launch made China the third country to independently launch a person into space, and represented the culmination of the efforts of the Chinese manned space program that began 11 years earlier. During the mission, taikonaut, People's Liberation Army Lieutenant (PLA) Colonel, Yang Liwei, made 14 orbits and landed just 21 hours after launch, only 5 km from the landing site in Inner Mongolia. Due to government fears that an unsuccessful mission would cause

Figure 8.1 Shenzhou-5 was China's first human spaceflight mission launched on October 15th, 2003. The Shenzhou spacecraft was launched on a Long March 2F rocket booster. There had been four previous flights of unmanned Shenzhou missions since 1999. Courtesy Wikimedia (*see colour section*).

Table 8.1. Shenzhou 5 mission parameters.

Mass	7,790 kg	Launch date	October 15, 2003 (01:00:03 UTC)
Perigee	332 km	Landing date	October 15, 2003 (22:22:48 UTC)
Apogee	336 km	Mission duration	21 hr, 22 min, 45 sec
Inclination	42.2°	Number of orbits	14
Period	91.2 min	Launch pad	Jiuquan Satellite Launch Center
Crew size	1	Taikonaut	Yang Liwei

embarrassment, neither the launch nor the re-entry was televised live, although the subsequent success of the mission was announced within minutes of each event!

Shenzhou 5 crew

Born in Suizhong, Liaoning Province on June 21st, 1965, PLA Air Force (PLAAF) pilot and China's first citizen in space, Yang Liwei, began his astronaut career upon being selected in China Group 1 in 1998, following a rigorous selection process described in Panel 8.1.

Panel 8.1. Taikonaut screening process

Selection of taikonauts to fly the Shenzhou spacecraft began at the end of 1995. Only PLAAF pilots were considered. Following a review of service records, 1,504 candidates were identified. This number was reduced to 866 following more stringent screening and, in the summer of 1996, 60 candidates passed initial testing at their home bases before being sent to Beijing for final aptitude tests and interviews, which whittled the numbers down to just 20 by April, 1997. From the group of 20, 12 taikonauts were selected at the end of 1997 and the group was officially established in January, 1998.

Liwei joined the PLAAF in September, 1983, and entered the No. 8 Aviation College, graduating as a fighter pilot in 1987 with the equivalent of a Bachelor's degree. By the time of his first spaceflight in 2003, Liwei had accumulated 1,350 flight hours. As one of 12 taikonauts, Liwei underwent five years of arduous physical, psychological, and technical training at the Astronaut Training Base (ATB) in Beijing. There, he received lessons in aviation dynamics, geophysics, meteorology, astronomy, space navigation, design principles of spacecraft, and systems training in space flight simulators. In addition to the theoretical instruction, the taikonauts also learned arctic, desert, and sea survival skills in case their capsule landed far from the recovery area. On September 20th, 2003, the group of taikonauts began training in the actual

Shenzhou 5 spacecraft at the JSLC. Unlike the American system, in which flight crews are announced as much as 18 months in advance of a mission, Liwei and the other taikonauts didn't know who would be selected just one month before the flight! In a process akin to *American Idol*, taikonauts with the lowest performance scores were deselected until finally, only three candidates remained, one of whom was Liwei.

Shenzhou 5 flight

China's first manned spaceflight began under a clear blue sky with the flawless launch of the LM-2F booster. Less than 10 minutes following launch, Shenzhou 5 entered an initial 200×343 km orbit. The mission plan (Table 8.2) was conservative, requiring Liwei to remain in the re-entry capsule for the duration of the flight. Thanks to four Chinese tracking ships deployed in oceans around the world, frequent communications were possible, including color television links. As Shenzhou entered its 21st orbit, the orbital module separated and retrofire was commanded via a tracking ship stationed in the Atlantic.

Political outcome

> "To establish myself as a fully trained astronaut, I have studied hard in my college years and have received training much tougher than for a fighter pilot. In order to achieve our goals, we must have the determination to face difficulties and to overcome them. I was pushed to my limits over five years of severe practical training that included tests involving G-forces and wilderness survival training. I am proud of becoming a member of the manned space flight project team and being able to witness the development of my country's space technology and industry. I hope my experience will encourage more people to become interested in space technology and support space development."
>
> Yang Liwei, following his historic flight

The mission was widely hailed as a triumph for Chinese science and technology and a milestone for Chinese nationalism, although Yang Liwei showed the flags of the United Nations in addition to the PRC flag.

> "An honor for our great motherland, an indicator for the initial victory of the country's first manned space flight and for an historic step taken by the Chinese people in their endeavour to surmount the peak of the world's science and technology."
>
> President Hu Jintao's remarks during an official celebration of Shenzhou 5

While continuing spaceflight training, Yang has also taken on new administrative duties as Deputy Director of the China Astronaut Research and Training Center and also Deputy Director General of China's manned space program. He earns a monthly salary of 10,000 yuan ($1,200).

Table 8.2. Shenzhou 5 flight chronology.

Time	*Event*
T – 7 days	October 8: Announcement made that launch is scheduled for October 15 and the flight will be made by one taikonaut
T – 2 days	October 13: Reports state that launch will take place between October 15 and 17. Hong Kong press sources incorrectly state that Li Qinglong will be the first Chinese taikonaut
T – 1 day	October 14: Yang Liwei is named as the first Chinese citizen in space. It is announced that the launch will not be televised live
Lift-off	October 15: 0100 GMT, 09:00 Beijing time
T + 00 hr 10 min	Shenzhou 5 enters 200 × 343-km orbit
T + 01 hr 30 min	Liwei begins first three-hour rest period
T + 06 hr 57 min	Shenzhou 5 circularizes orbit to 343 km
T + 08 hr 30 min	Liwei conducts communication session with General Cao Gangchuan, China's Defence Minister
T + 11 hr 00 min	Liwei communicates with his wife and son. Live video is relayed from the capsule showing Liwei displaying Chinese and United Nations flags
T + 14 hr 00 min	Liwei beings second three-hour rest period
T + 20 hr 35 min	Command radioed to initiate re-entry
T + 20 hr 36 min	Orbital module separates and remains in 343-km circular orbit
T + 20 hr 38 min	Retrofire begins
T + 20 hr 59 min	Service module separates from re-entry capsule
T + 21 hr 00 min	Re-entry capsule enters Chinese territory
T + 21 hr 04 min	Re-entry capsule enters communications blackout due to sheath of ionized air around capsule
T + 21 hr 07 min	Re-entry capsule exits communications black-out. Recovery helicop ters receive signal enabling them to estimate landing point
T + 21 hr 23 min	Re-entry capsule soft-lands 4.8 km from intended landing point and 7.5 km from recovery vehicles
T + 21 hr 28 min	Capsule sighted by recovery team
T + 21 hr 36 min	Recovery team arrives at capsule
T + 21 hr 51 min	Liwei exits capsule
T + 22 hr 40 min	Following medical examination, Liwei returns to Beijing

Shenzhou 6

China's second manned spaceflight (Table 8.3), Shenzhou 6, was launched on October 16th, 2005, on an LM-2F launch vehicle from the JSLC. The Shenzhou spacecraft carried taikonauts Fèi Jùnlóng (Commander) and Niè Hǎishèng (Flight Engineer) for nearly five days in low Earth orbit (LEO), during which the two spacefarers conducted scientific experiments and engaged in various media activities.

Table 8.3. Shenzhou 6 mission parameters.

Mass	8,040 kg	Launch date	October 12, 2005 (01:00:03 UTC)
Perigee	334 km	Landing date	October 16, 2005 (20:33:00 UTC)
Apogee	334 km	Mission duration	4 days 19 hr 33 min
Inclination	42.4°	Number of orbits	75
Period	91.2 min	Launch pad	Jiuquan Satellite Launch Center
Crew size	2	Taikonauts	Fèi Jùnlóng and Niè Hǎishèng

Shenzhou 6 Crew

Fèi Jùnlóng and Niè Hǎishèng were chosen from a pool of five pairs of taikonauts. One month prior to the flight, the two pairs with the lowest performance were deselected, leaving three pairs eligible. A similar process had been followed prior to the flight of Shenzhou 5, during which Niè Hǎishèng and eventual Shenzhou 6 back-up crewmember Zhái Zhìgāng had been in the final group.

Born May, 1966, Fèi Jùnlóng is a graduate of the No. 9 Aviation School and Changchun No. 1 Flight College of the PLAAF. After graduating from the Flight Training School of the Air Force with excellent marks, Jùnlóng worked as a PLAAF pilot, flight instructor, and flight technology inspector before being selected as an astronaut in 1998.

Fellow taikonaut, Niè Hǎishèng, was born in Zaoyang, Hubei, on October 13th, 1964. After completing flight training with the PLAAF in 1984, Hǎishèng began service as a fighter pilot in June, 1989, before serving as commander of a squadron. In common with Jùnlóng, Hǎishèng entered active space service when he joined China Group 1 in 1998. By the time of his space flight, he had accumulated 1,480 flying hours.

Shenzhou 6 Flight

Reportedly costing $110 million (900 million yuan), Shenzhou 6 was a significantly improved version of its predecessor, with more than 100 modifications, such as a better environmental control life support system (ECLSS) and updated computers. Following a problem-free launch and orbital insertion, Jùnlóng and Hǎishèng entered the orbital module while remaining in their flight suits. After checking pressure integrity, the crew changed into blue intravehicular activity (IVA) suits and activated experiments in the orbital module. Throughout the mission, the crew slept in shifts, with one crewmember awake in the descent module and the other asleep in the orbital module.

On the mission's second day, the crew tested the spacecraft's automatic motion damping system by purposefully using violent movements while donning and doffing

Table 8.4. Shenzhou 6 re-entry chronology.

Time (GMT)	*Event*
19:42	Tracking vessel, *Yuanwang 3*, acquires Shenzhou 6
19:43	*Yuanwang 3* receives verification Shenzhou is oriented for retrofire
19:44	Orbital module separation
19:45	Retrofire. Shenzhou 6's main engines ignite
19:48	Main engine shutdown
20:07	Service module separates from re-entry capsule
20:20	Re-entry capsule completes aerodynamic braking. Main parachute deployed at subsonic speed
20:23	Re-entry capsule's heat shield jettisoned
20:32	Soft landing within sight of recovery forces

their space suits. Later in the day, Shenzhou 6's engines were operated for a few seconds to trim the orbit from a circular orbit to a slightly elliptical one.

During the fourth day, the crew conducted various space science experiments, tested different space suit models, and spoke with Chinese President, Hu Jintao, before making preparations for their return to Earth. On their final day in orbit, the crew prepared Shenzhou 6 for landing and donned their space suits for re-entry (Table 8.4).

Political outcome

China described its second manned spaceflight as a perfect success that would not only improve future space exploration efforts, but also enhance national unity. Speaking hours after the landing, Tang Xianming, Director of the Manned Space Engineering Office, stated that the success of Shenzhou 6 demonstrated China was among the best in the world in science and technology:

> "We believe this great achievement will inspire greater patriotic passion, national pride and cohesion, stimulate our enthusiasm for scientific exploration and originality, advance a giant leap in science for our country, improve China's overall national strength, and strive for the revival and prosperity for the whole Chinese nation."
>
> Tang Xianming, Director of the Manned Space Engineering Office

Echoing Xianming's statement were the comments made by Wu Bangguo, China's top legislator, who was quoted as saying that Shenzhou 6's success would improve China's international status and national strength and help mobilize people around the Communist Party. Other post-flight announcements indicated the next Shenzhou flight would feature China's first extravehicular activity (EVA). This flight would be

Table 8.5. Shenzhou 7 mission parameters.

Mass	8,040 kg	Launch date	September 25, 2008 (13:10:04 UTC)
Perigee	330 km	Landing date	September 28, 2008 (09:37:00 UTC)
Apogee	336 km	Mission duration	2 days 20 hr 27 min
Inclination	42.4°	Number of orbits	45
Period	91.2 min	Launch pad	Jiuquan Satellite Launch Center
Crew size	3	Taikonauts	Zhái Zhìgāng, Lìu Bó Mìng, Jing Haipeng

followed by an unmanned rendezvous and docking (R&D) mission, which in turn would be followed by Shenzhou 10, provisionally scheduled for 2012, which would demonstrate a *manned* R&D capability.

Shenzhou 7

Launched almost a year behind schedule, Shenzhou 7 (Table 8.5) was China's third spaceflight. Its primary goal was to conduct China's first EVA, which, if successful, would make China the third country, after Russia and the US, capable of conducting EVA operations.

Shenzhou 7 crew

Born in Longjiang County, Qiqihar, Heilongjiang province, October 10th, 1966, Zhái Zhìgāng (Figure 8.2) enrolled at the PLAAF Aviation Institute and studied to become a fighter pilot. After serving as a squadron leader and pilot trainer, Zhìgāng was selected for the taikonaut program in 1998 and was one of three candidates of the final group to train for Shenzhou 5 and also one of the six taikonauts in the running for selection for Shenzhou 6.

Lìu Bó Mìng, born in September, 1966, in Yi'an County of China's northeast Heilongjiang province, served as a fighter pilot in the PLAAF before being selected as a taikonaut in 1998. During his first space mission, Lìu wore a Russian Orlan-M space suit, remaining in the hatch of the orbital module, assisting Zhìgāng in performing his EVA.

The third member of the Shenzhou 7 mission, Jĭng Hăipéng, was born on October 24th, 1966. Like Zhìgāng and Lìu, Hăipéng served as a fighter pilot with the PLAAF until being selected as a taikonaut in 1998, and, like Zhìgāng, he was one of the six candidates considered for the Shenzhou 6 mission.

Figure 8.2 Zhái Zhìgāng. Courtesy CNSA.

Flight

To support the mission's (Table 8.6) primary objective of conducting an EVA, the China Academy of Space Technology (CAST) developed the Feitian EVA space suit. The suit was derived from the technology of the Russian Orlan space suit designed and developed by NPP Zvezda. Both the Orlan and the Feitian space suits were used during the EVA, which took place on September 27th. In preparation for the EVA, Zhìgāng and Bó Mìng entered the orbital module on September 26th to begin preparation and assembly of the space suits – a process that took about 10 hours. Once the safety checks had been completed, Zhìgāng and Bó Mìng began preparation for the EVA while Haipeng remained in the re-entry module to monitor the spacecraft. At this point, the hatch between the orbital module and the re-entry module was sealed. Following confirmation from ground control that Zhìgāng was the EVA taikonaut and Bó Mìng was the supporting taikonaut, the two crewmembers assisted one another with donning the space suits and conducting systems checks. Once they had donned their space suits, the orbital module was depressurized and

Table 8.6. Shenzhou 7 flight chronology.

Time	*Event*
T – 77 days	Shenzhou 7 airlifted from Beijing to Jiuquan Satellite Launch Centre (JLSC)
T – 50 days	ChangZheng 2F launch vehicle arrives at JLSC by rail
T – 7 days	Integration of CZ-2F and Shenzhou spacecraft completed
T – 5 days	CZ-2F/Shenzhou 7 spacecraft moved from vertical launch vehicle processing facility to launch pad
Lift-off	Shenzhou 7 launched from JSLC at 21:10 local time
T + 13 hr 10 min	Zhìgāng and Bó Mìng begin assembly of EVA suits
T + 1 day 00 hr 50 min	Preparations and safety checks of EVA suits completed
T + 1 day 15 hr 18 min	Zhìgāng and Bó Mìng enter orbital module to prepare for EVA mission. Hatch between re-entry and orbital module is sealed
T + 1 day 18 hr 19 min	Zhìgāng and Bó Mìng complete donning of space suits and begin depressurization of orbital module
T + 1 day 19 hr 21 min	Zhìgāng begins EVA operation
T + 1 day 19 hr 46 min	Zhìgāng completes EVA operation
T + 1 day 22 hr 14 min	Macro satellite launched
T + 2 day 17 hr 50 min	Crew don their pressure suits and begin preparations for re-entry
T + 2 day 20 hr 04 min	Service module separates from re-entry capsule
T + 2 day 20 hr 19 min	Re-entry capsule deploys main parachute
T + 2 day 20 hr 28 min	Re-entry capsule touches down

oxygen and power supplies on the space suits were switched to autonomous working mode. Although some minor problems were experienced with the orbital module's exterior hatch, Zhìgāng's egress from the orbital module was uneventful. Following an EVA lasting 25 minutes and 23 seconds, during which Zhìgāng waved a Chinese national flag to the external camera onboard the spacecraft, Zhìgāng re-entered the orbital module with Bó Mìng's assistance. Once the orbital module's hatch was sealed and the module re-pressurized, the taikonauts returned to the re-entry capsule.

Having performed the primary mission objective, the taikonauts turned their attention to the secondary objective of launching a macro satellite carried on top of the orbital module. The 40-kg satellite, equipped with cameras, captured pictures of Shenzhou 7 as it flew in formation with the mother ship, before gradually moving away under control of a ground tracking station. Following the taikonauts' return to Earth, the macro satellite performed simulated R&D maneuvers with the orbital module, providing mission planners with valuable experience for future docking operations.

The third and final objective of Shenzhou 7 was to perform testing on a satellite data relay system that provided data relay services between the spacecraft and the mission ground station.

In addition to the media coverage of China's first EVA, Shenzhou 7 also attracted attention when its macro satellite passed within 45 km of the International Space

Station (ISS) on September 27th. China did not respond when asked why it allowed the satellite to pass so close to the ISS. However, several newspapers noted China's track record of using its Shenzhou missions for dual-military purposes, and speculated the close encounter with the ISS may have been used as an opportunity to test anti-satellite (ASAT) interception technology.

The budding Chinese space program, which came of age in 2003 with its first manned launch, has deservedly received widespread media attention, much of it suggesting a new space race is imminent. Encouraged by their achievements, the Chinese have now set their sights on the Moon, but it won't be easy. The US required only eight years to progress from their first manned space flight to the first lunar landing – a triumph that was achieved while simultaneously developing the technology to do it. Could a committed nation such as China achieve such a goal? While China undoubtedly belongs in the top tier of spacefaring nations by virtue of its manned spaceflight capability, its lack of expertise in this arena compared with the US means that it faces a challenge even greater than the one faced by the Americans nearly 50 years ago. To illustrate the chasm between the manned spaceflight expertise of the two competitors in the new space race, one needs only to briefly review the tremendous depth and breadth of US spacefaring experience.

US MANNED SPACEFLIGHT EXPERTISE

Since NASA's Mercury astronauts made their first brief forays into space more than 40 years ago, nearly 300 US astronauts have traveled into orbit. But, whereas the Mercury flights lasted only a matter of hours, NASA now sends its astronauts to the ISS for mission increments lasting up to six months. Four decades after the dawn of the manned spaceflight era, NASA is not only extending the range and sophistication of human operations in space, but also learning how to live and stay in space. To really appreciate the extent of the US's manned spaceflight experience, it is appropriate to review some of the highlights.

A brief history of notable spaceflights

Project Mercury

The first American astronauts, in common with the new cadre of taikonauts, were drawn from the ranks of fighter pilots. On May 5th, 1961, America's first manned spaceflight took place with the launch of Alan B. Shepard Jr, from Cape Canaveral on a Mercury Redstone launch vehicle to an altitude of 87 km in a suborbital flight lasting less than 15 minutes. Shepard's mission was followed by John Glenn's flight onboard *Friendship 7* on February 20th, 1962, when Glenn became the first American to circle the Earth. The final flight of Project Mercury took place between May 15th and 16th, 1963, with the launch of Gordon Cooper, who orbited the Earth 22 times in a mission lasting 34 hours.

Project Gemini

While Project Gemini didn't have the excitement of Apollo, its success was critical to Kennedy's goal of reaching the Moon before the end of the decade. Gemini's primary objective was to demonstrate R&D techniques that would be used during the Apollo missions, when the lunar lander would separate from the command module while in lunar orbit. Another objective of the program was to extend the time spent in space to two weeks.

Less than two years following the end of Project Mercury, the first operational mission of Project Gemini took place on March 23rd, 1965, with the launch of Mercury astronaut Gus Grissom, and naval aviator, John Young. The second manned Gemini mission followed less than four months later – a mission notable for the first American EVA, performed by Edward H. White III. Perhaps more notable than White's EVA was the routine manner in which NASA launched astronauts, as evidenced by the program's 10 manned missions in less than 20 months. Such an aggressive flight rate enabled 16 new astronauts to chalk up nearly 1,000 hours of space experience in the years between Mercury and Apollo, which, by 1966, was approaching flight readiness.

Apollo Program

The launch of the first manned flight of the Apollo Program (Apollo 7) took place on October 11th, when Walter M. Schirra Jr, Donn F. Eisele, and Walter Cunningham took off atop a Saturn 1B launch vehicle. Apollo 8 followed less than two months later with the launch of Frank Borman, James A. Lovell Jr, and William A. Anders atop a Saturn V booster from the Kennedy Space Center (KSC). Initially, the mission was planned as a test for Apollo hardware in LEO, but senior engineer George M. Low of the Manned Spacecraft Center (renamed the Johnson Space Center in 1973) and Apollo Program Manager, Samuel C. Philips, sought approval for an ambitious circumlunar flight, which was authorized in November. Following 1.5 Earth orbits, Apollo 8's third stage performed a trans-lunar insertion (TLI) burn, placing the spacecraft on a trajectory for the Moon. As it traveled towards the Moon, the crew focused a portable television camera towards the Earth, returning epochal images of the Earth hanging in the blackness of space.

Apollo 11, the first lunar landing mission, lifted off on July 16th, 1969. Following systems checks to confirm all systems were functioning normally, the crew of Neil A. Armstrong, Edwin E. Aldrin, and Michael Collins began the three-day trip to the Moon. On July 20th, at 4:18 p.m. EST, the Lunar Module, occupied by Armstrong and Aldrin, landed on the lunar surface while Collins remained in lunar orbit. After delivering his famous "one small step" speech, Armstrong was joined by Aldrin on the surface and together they lumbered around the landing site in the one-sixth lunar gravity, planting an American flag and collecting rock samples. The following day, they launched to lunar orbit, docked with the Apollo command module and returned to Earth, splashing down in the Pacific on July 24th.

One of the near disasters of the Apollo Program occurred in April, 1970, with the flight of Apollo 13, when, less than three days into the flight, an oxygen tank in the Apollo service module ruptured, damaging power and life support systems. Using the lunar module as a lifeboat and by employing some now legendary NASA ingenuity, the crew of James A. Lovell, Jr, Fred W. Haise, Jr, and John L. Swigert, Jr, returned safely on April 17th, 1970, the near disaster solidifying in the minds of the public NASA's technological brilliance.

The first of the longer, expedition-class lunar missions was launched on July 26th, 1971. Apollo 15 was the first mission to use the lunar rover to extend the astronauts' surface exploration activities. Astronauts David R. Scott and James B. Irwin rode more than 27 km in the rover and brought back a sample of ancient lunar crust nicknamed the *Genesis Rock*, one of the prize trophies of the Apollo Program.

Less than 18 months later, Apollo 17, the final mission of the program following the cancellation of missions Apollo 18, 19, and 20, returned from the Moon. The manned spaceflight experience NASA gained during Apollo is incalculable. Apollo astronauts logged some 84,000 hours (nearly 10 man-years) preparing for their missions.

Skylab

NASA had studied several concepts for space stations but it wasn't until the development of the mighty Saturn V in the mid-1960s that the Skylab Program could be realized. Skylab was born with the twin objectives of using leftover Apollo hardware and achieving extended stays in LEO. Divided into two levels separated by a metal grid floor, Skylab (Figure 8.3) had a habitable volume equal to a three-bedroom house, and featured many of the comforts of home, including a dining table, a work area, a shower, and a bathroom.

Launched on May 14th, 1973, America's first space station was occupied by three different crews over the next nine months, for one, two, and then three months at a time. Orbiting at an altitude of 434 km, the space station was deactivated between flights. The nine Skylab astronauts clocked 513 days in orbit, demonstrating not only the value of having humans working for extended durations in space, but also that astronauts could survive the experience.

Apollo–Soyuz Test Project

Taking place at the height of the détente between the US and the Soviet Union, the Apollo–Soyuz Test Project (ASTP) was launched on July 15th, 1975. The mission was conducted using existing Apollo and Soyuz spacecraft, the Apollo vehicle being almost identical to the one that orbited the Moon, and the Soviet spacecraft being the same one as used for flying cosmonauts since its introduction in 1967. Although the flight was more a symbol of the lessening tensions between the two superpowers than a genuine scientific endeavor, the mission also tested the compatibility of

Figure 8.3 Skylab. Courtesy NASA.

American and Soyuz R&D systems, presumably to pave the way for international space rescue capabilities.

Space Shuttle Program

Before the Space Shuttle, launching payloads into space was a one-way proposition because satellites placed in LEO could not be returned to Earth. Approved as a national program in 1972, the Space Shuttle not only revolutionized the delivery of payloads to LEO, it also radically changed the way astronauts worked in space. Part spacecraft and part aircraft, the development of the Space Shuttle required several innovative technological advances and highly complex engines capable of being reused repeatedly.

In terms of manned spaceflight, the Space Shuttle, capable of carrying as many as eight astronauts, changed the sociology of space travel with the creation of new classes of astronauts such as mission specialists and payload specialists, who were selected following a rigorous selection process (Panel 8.2).

Panel 8.2. NASA astronaut selection process

NASA's Astronaut Candidate selection process considers both civilian and military personnel who meet a series of minimum requirements. For non-pilots, potential Astronaut Candidates must hold a Bachelor's degree, followed by at least three years of related professional experience. Additional qualifications include the ability to pass NASA's long-duration spaceflight medical. For those applying to become a Pilot Astronaut, requirements include a Bachelor's degree in engineering, science, or mathematics, at least 1,000 hours pilot-in-command time in jet aircraft, and the ability to pass NASA's long-duration spaceflight physical. Those applicants selected for training as an Astronaut Candidate undergo a training and evaluation period lasting approximately two years. Following completion of training, astronauts are eligible for flight assignments, usually requiring two to three years' additional training.

Highlights of the Space Shuttle Program

April 12th, 1981: The reusable Shuttle fleet debuted on April 12th, 1981, with the launch of the Space Shuttle *Columbia* (STS-1), crewed by John Young and Bob Crippen (Figure 8.4). In LEO, Young and Crippen tested *Columbia*'s systems, fired the Orbital Maneuvering System (OMS) and the Reaction Control System (RCS), and opened and closed the payload bay doors, before making a smooth touchdown at Edwards Air Force Base (EAFB) after completing 36 orbits.

June 18th, 1983: The first Shuttle flight with five crewmembers was STS-7, which also carried America's first female astronaut, mission specialist, Sally K. Ride. During the mission, piloted by Bob Crippen and Frederick H. Hauck, the crew launched two communications satellites and the reusable Shuttle Pallet Satellite (SPAS).

November 28th, 1983: STS-9, piloted by John Young and Brewster Shaw, carried West German astronaut, Ulf Merbold, the first non-US astronaut to fly in the American space program. The flight was also notable for the first flight of Spacelab 1, a space laboratory carrying more than 70 experiments in areas of life sciences, space plasma physics, solar physics, Earth observations, and materials science.

Figure 8.4 A dramatic evening photo of *Columbia* sitting on the pad at Kennedy Space Center. Courtesy NASA (*see colour section*).

January 28th, 1986: One of the most significant events of the 1980s was the tragedy of the Space Shuttle *Challenger*, which exploded just 73 seconds after launch as a result of a leak in one of the solid rocket boosters (SRBs) that ignited the external tank (ET) carrying liquid fuel. The disaster resulted in a thorough review of the safety of the Shuttle Program, leading to several reforms in the management structure and manned spaceflight procedures.

April 24th–29th, 1990: During the flight of STS-31, the Hubble Space Telescope (HST) was deployed, but the mission was overshadowed by the discovery of a spherical aberration preventing the telescope from focusing all light to a single point. Although NASA received negative publicity, engineers planned servicing missions to correct it.

December 2nd–12th, 1993: The HST servicing mission was STS-61. Piloted by Richard Covey and Kenneth Bowersox, the Space Shuttle *Endeavour* performed a rendezvous with the stricken telescope and, following a precise grappling maneuver, the HST was berthed in the Shuttle's cargo bay. There, the flight crew, working with ground controllers in Johnson Space Center (JSC), completed 11 servicing tasks during five EVAs, restoring the HST to a full bill of health. The HST was then re-boosted into its orbit.

February 3rd–11th, 1994: STS-60 marked the beginning of the Shuttle–Mir era, with an historic mission featuring the flight of Russian cosmonaut, Sergei Krikalev, onboard the Space Shuttle *Discovery*. The Shuttle–Mir Program was a collaborative program between the US and Russia, involving cosmonauts flying onboard the Space Shuttle, visits by the Space Shuttle to the Russian space station, Mir, and American astronauts flying onboard Soyuz spacecraft to engage in long-duration increments onboard Mir. In addition to underlining the new cooperation between the US and Russia, STS-60 also highlighted Russia becoming a partner in the development of the ISS.

February 3rd–11th, 1995: One year after the groundbreaking flight of STS-60, *Discovery* flew another notable mission (STS-63), when the Orbiter, flown for the first time by a female pilot, Eileen Collins, conducted a flyby of Mir.

September 16th–26th, 1996: STS-79, the fourth Shuttle–Mir docking mission, featured the return of NASA astronaut, Shannon Lucid, after 188 days in orbit and the first US crew exchange aboard Mir. Lucid's long-duration mission set a new US record for the longest stay in space.

September 25th–October 6th, 1997: STS-86, the seventh Shuttle–Mir flight, continued the presence of a US astronaut onboard Mir with the transfer of David Wolf, who became the sixth NASA astronaut in succession to live on Mir. STS-86 was also notable for the first joint US–Russian EVA during a Shuttle mission and returning Mike Foale. After spending 145 days in space, Mike Foale returned to Earth onboard *Atlantis*, having clocked an estimated 93,000,000 km in space, making his the second longest US spaceflight, after Lucid's record of 188 days.

December 4th–15th, 1998: STS-88 marked the beginning of the assembly of the ISS. One of the primary mission objectives was the mating of the Zarya control module with the Unity connecting module – a task successfully completed following three EVAs. Other EVA objectives included the testing of a Simplified Aid for EVA Rescue (SAFER) self-rescue device for an astronaut who might become separated from the spacecraft during an EVA, deploying antennae on Zarya into position, and installing a handrail on Zarya to help future astronauts during EVA activities. During the mission, Jerry Ross established a new spacewalk record with seven EVAs totaling 44 hours 9 minutes.

April 8th–19th, 2002: The launch of STS-102 marked another milestone for Jerry Ross, who became the first human to fly in space seven times. During the flight, he performed two EVAs, increasing his total time outside the confines of a spacecraft to more than 58 hours – a duration only surpassed by cosmonaut, Anatoly Solovyev. The primary objective of STS-102 was the installation of an ISS truss containing navigational devices, computers, and power systems needed by the additional laboratories of the ISS. The truss, which had been carried inside *Atlantis*'s payload bay was lifted out of the payload bay using the Canadarm 2, and maneuvered onto a clamp on the top of the Destiny Lab module of the ISS.

Launch date: January 16th, 2003: After a 16-day mission, *Columbia* and her crew were scheduled to land at KSC on February 1st, but, 16 minutes prior to planned touchdown, the Orbiter disintegrated during re-entry over East Texas, resulting in loss of the vehicle and her crew. The tragedy occurred as a result of the Orbiter's vital heat shield having been damaged during lift-off, allowing superheated atmospheric gases to enter a hole in the vehicle's left-wing edge. A team of NASA astronauts, engineers, and pilots spent more than four years reconstructing the descent from orbit of *Columbia* and her crew in a report entitled the *Columbia Crew Survival Investigation Report*, which made 30 recommendations to enhance astronaut safety and survivability in the event of a future catastrophic event.

July 26th–August 9th, 2005: STS-114 marked the *return to flight* mission of the Space Shuttle program with the launch of *Discovery*, ending a 2.5-year hiatus in Shuttle operations. As a result of the fate that befell *Columbia*, *Discovery*'s flight was designed with a back-up plan that would require the crew to take refuge onboard the ISS in the event of a problem preventing re-entry. Mission managers would then consider the possibility of mounting a risky rescue mission using another Shuttle. Fortunately, except for weather issues forcing a landing at EAFB, *Discovery*'s mission was flawless, and was notable for the implementation of several safety improvements, such as sensors embedded in the Shuttle's wings, new foam applications on the ET, and upgraded cameras. STS-114 also featured the first Rendezvous Pitch Maneuver, which flipped the Orbiter end over end, allowing ISS crewmembers to photograph the Orbiter's underside and its heat-resistant tiles. In addition to aerobatic maneuvers, Shuttle crewmembers performed EVAs to test the Shuttle's new tile repair system, and installed new gyroscopes on the ISS.

December 12th–22nd, 2006: One of the most challenging Shuttle missions took place at the end of 2006 with the launch of *Discovery*. STS-116, the 20th Shuttle flight to the ISS, was a 13-day mission tasked with rewiring the station's power system and installing an integrated truss segment. The mission also witnessed the transfer of NASA astronaut, Sunita Williams, who replaced European Space Agency (ESA) astronaut, Thomas Reiter, who had served on the ISS since July, 2006. The mission added to NASA's already considerable EVA experience as a result of four EVAs being required to complete all the electrical tasks, and yet another Shuttle EVA record was set by Curbeam, who, as the only STS-116 crewmember to participate in all four EVAs, set a record for the most EVAs performed by one astronaut during a single mission.

International Space Station 10th anniversary

On November 20th, 2008, nations around the world joined together to mark a major milestone in space exploration, celebrating the 10th anniversary of the largest spacecraft ever built. The product of more than 30 construction flights and nearly 150 EVAs, the ISS represents an international construction project of unprecedented complexity and sophistication. Equally impressive was that the project had involved international cooperation between NASA, the Russian Space Agency, the Canadian Space Agency (CSA), the Japanese Aerospace Exploration Agency (JAXA), and 11 members of the European Space Agency (ESA). By the time of the station's 10th anniversary, the ISS had completed 57,309 orbits, covering a distance of 2,292,360,000 km.

Anatomy of a Space Shuttle mission: STS-116

STS-116 (Table 8.7), the most complex ISS construction mission to date, began with the rollout of the Shuttle *Discovery* to Pad 39B on November 9th. Covering the 6.7-km distance from NASA's cavernous 52-story Vehicle Assembly Building (VAB) in a little more than seven hours while fixed to the crawler carrier vehicle, the 100-tonne *Discovery* carried an assorted cargo of ISS tools, spare parts, and the all-important Port 5 (P5) Truss.

Three weeks later, on December 1st, *Discovery*'s crew commenced their quarantine at JSC's Quarantine Facility to stay healthy before their planned 12-day flight. While at the facility, the crew made a ground-call to their fellow spacefarers onboard the ISS to prepare for the station assembly mission. On

Table 8.7. STS-116.

Mission overview			
Launch	December 9, 2006, 8:47 p.m. EDT	Mission duration	12 day, 20 hr, 45 min
Landing	December 22, 2006, 5:32 p.m. EDT	Landing site	Kennedy Space Center
Orbiter	*Discovery*	Inclination/altitude	51.6°/215 km
Mission number	STS-116	Primary payload	P5 Truss, SPACEHAB
Launch pad	39B	Shuttle lift-off weight	2,055,159 kg
Crew			
Commander	Mark L. Polansky	Pilot	William A. Oefelein
Mission specialists: Joan E. Higginbotham, Robert L. Curbeam, Nicholas J.M. Patrick, Sunita L. Williams, Christer Fuglesang (ESA)			

December 4th, NASA began the three-day countdown to the launch. During the countdown, ISS controllers boosted the station into a higher orbit in preparation for docking with *Discovery*. On the same day, *Discovery*'s crew arrived at JSC, jetting in on T-38 training aircraft. Still under quarantine, Commander Mark Polansky and Pilot William Oefelein practiced shuttle landings as part of their final pre-launch preparations.

On December 6th, following last-minute technical issues involving a brief power surge in circuits responsible for transferring power from the mobile launch platform to *Discovery*, NASA mission managers gave the go-ahead for launch. The next day, Shuttle workers unveiled *Discovery*, rolling back the protective Rotating Service Structure (RSS), and began the three-hour fuelling of the Orbiter's ET with 500,000 gallons of super-cold liquid hydrogen and liquid oxygen. Shortly after commencement of fuelling, *Discovery*'s crew attended a weather briefing and ate their last meal at Crew Quarters before suiting up and leaving for the launch pad. In the late afternoon, the astronauts entered the orbiter and the hatch was closed at 7:30 p.m. EST. With launch just two hours away, the crew busied themselves with checklists and configuring the cockpit for the spacecraft's ascent to orbit, but, due to a thick layer of low clouds, the launch was scrubbed.

Two days later, clad in their "Day-Glo" orange launch and entry suits, the crew went through the same prelaunch ritual, listening to weather reports from the primary abort sites and information about the crosswinds at the Shuttle Landing Facility (SLF). With reports indicating acceptable limits, the Orbiter's access arm retracted, the liquid hydrogen and liquid oxygen tanks pressurized, and *Discovery*'s flight computers took control of the countdown. At 6.6 seconds before take-off, the three Space Shuttle main engines ignited and, at T – 0, the twin SRBs lit up the night sky, sending *Discovery* on her fiery ascent to orbit. Two minutes and three seconds after launch, the SRBs separated, falling back to the Atlantic, to be recovered. Eight minutes after launch, the ET shut down and was discarded. Twenty-two minutes after launch, *Discovery* reached orbit and the seven STS-116 astronauts prepared for the first post-launch news conference, before preparing for their mission tasks (Table 8.8).

The second day of the mission was spent inspecting *Discovery*'s thermal protection system (TPS), using the Orbiter's 15-m robotic arm and a sensor-laden extension boom and preparing for the R&D operations with the ISS the following day.

The highlight of *Discovery*'s third day in orbit was the Orbiter's approach to the ISS, which included an R&D procedure including a back-flip pirouette maneuver to permit station crewmembers to take digital images of the Shuttle's TPS. At approximately 180 m from the station, *Discovery* performed a circular pitch-around below the station. The nine-minute flip enabled Expedition 14 Commander Mike Lopez-Alegria and Flight Engineer Mikhail Tyurin to document the required imagery of the TPS before transmitting the images to the Mission Evaluation Room at JSC. Following the back-flip maneuver, Polansky guided *Discovery* to the docking port of the ISS Destiny module. Shortly after docking, the crew entered the ISS, and Sunita Williams officially relieved ESA astronaut, Thomas Reiter, as flight engineer of Expedition 14.

Table 8.8. STS-116 mission timeline.

Flight Day 1 Launch Payload bay door open Canadarm power-up	*Flight Day 5* P4 SARJ activation	*Flight Day 9* Transfer work
Flight Day 2 Inspect Shuttle's TPS Space suit checkout and EVA preparation Airlock preparations	*Flight Day 6* EVA #2 ISS TVIS gyroscope replaced	*Flight Day 10* Final transfer work Undock and ISS fly-around Separation from ISS
Flight Day 3 Rendezvous operations RPM maneuver Dock with ISS	*Flight Day 7* Shuttle–ISS transfer work Crew off-duty time Crew news conference	*Flight Day 11* Cabin stowage RCS hot-fire test Flight control checkout
Flight Day 4 P5 spacer truss installed EVA #1	*Flight Day 8* EVA #3	*Flight Day 12* De-orbit preparations KSC landing

On December 13th, during the first EVA, the P5 Truss was aligned with the end of the P3/P4 Truss, extending the station's backbone by almost 4 m. The following day, while performing the mission's second EVA, the ISS's electrical system was reconfigured – a task completed during the mission's third EVA two days later. A fourth (unscheduled) EVA was performed on December 18th to resolve an issue with an uncooperative solar array, which needed to be retracted before the arrival of Russian Soyuz modules the following year. During an EVA lasting more than six hours, the troublesome array was resolved, and the Shuttle astronauts began their preparations for departure.

On December 19th, *Discovery* separated from the ISS to begin two days of homecoming preparations, including another check of the Orbiter's TPS, and a checkout of the landing systems.

Future of American manned spaceflight

There have been other periods in NASA's history when the agency has had no manned spaceflight capability, similar to the one facing it once the Shuttle retires in 2010. One such interlude was the hiatus following the *Challenger* tragedy in 1986 and again following the catastrophe that befell *Columbia* in 2003. However, the impending interval following the Shuttle's retirement in November, 2010, could become the longest hiatus in American manned spaceflight if the rollout of the agency's new family of launch vehicles is significantly delayed.

Although the media have mostly decried the interruption in American flights, the lull is no accident. The Bush Administration decided to retire the Shuttle fleet as a part of its plans to stop using the aging and increasingly risky Shuttles, before moving on to the next phase of manned spaceflight, as envisioned by the Vision for Space Exploration (VSE). The decision to relinquish the nation's access to space was in no small part driven by the *Columbia* tragedy and the need to divert funds from the Shuttle Program to the Constellation Program tasked with returning humans to the Moon and then sending them to Mars.

The critics are not the only ones voicing their concerns regarding the state of the nation's manned spaceflight capability. Former NASA Administrator, Dr Michael Griffin, often expressed his frustration with the choice of the Shuttle retirement date. In testimony to Congress, Griffin warned of the potential dire consequences of such action, suggesting a prolonged hiatus in manned spaceflight capability may see Chinese taikonauts setting foot on the Moon before American astronauts. Such a prospect is also a concern for Republican, Tom Feeney. Feeney has repeatedly urged his colleagues to take a stronger stand against decommissioning the Shuttle fleet and has warned of the repercussions of NASA's sitting back and allowing space capabilities to empower countries such as China, who do not have America's best interest at heart. Not all of his colleagues share Feeney's sense of urgency, however. For example, a fellow congressman recently suggested naming the new lunar base after Neil Armstrong – a suggestion to which Feeney responded by questioning why the congressman thought the Chinese would give the US permission to name a Chinese base after an American astronaut!

Polls have continually shown significant support for manned spaceflight. Combined with strong support from key members of Congress, manned spaceflight will continue, but with the US in the middle of a recession, space exploration will be viewed as a luxury rather than a necessity, and the NASA budget may become an easy target for cutbacks. In fact, with everything from housing to health care appearing to be in a state of crisis, getting man back to the Moon will be low on the list of President Obama's priorities. With China making huge leaps in manned spaceflight, the consequences of a manned spaceflight crisis at NASA may be grave. With day-to-day manned spaceflight for the US limited to Soyuz until *Orion* becomes operational, it may be half a decade or more before another NASA astronaut maneuvers a manned American spacecraft in LEO. The gap between the retirement of the Shuttle and *Orion* coming on stream will cause other problems for NASA, such as how to maintain an experienced cadre of astronauts when American manned spaceflight will consist of chaperoned flights onboard Soyuz capsules. Fortunately, thanks to the success of SpaceX, the anticipated five-year hiatus may be dramatically reduced and the hope is that by the time *Orion* is operational, NASA will have developed a safer launch system capable of returning American astronauts not just to orbit, but to the Moon and beyond. The Shuttle–*Orion* transition is necessary, but the management of the gap and China's manipulation of it may ultimately result in relegating the US to second place in the manned spaceflight business – a position echoing the space race with the Soviets in the late 1950s and early 1960s.

Ever since Yang Liwei's historic flight in 2003, the US space community has wondered how it might feel when one of the Apollo lunar landing flags is returned to Earth and displayed in a Beijing museum. For a nation that considers itself the only true space power, such a reality is not only unthinkable, but unimaginable. Simply put, the US will not entertain the possibility of a world in which the human spaceflight program of another nation (especially one flying a red flag!) is on the rise, while theirs is in decline. While some analysts wonder whether China's achievements mark the beginning of the end of the US dominance in space, in reality, China's three manned spaceflights and an EVA mark accomplishments achieved by the US more than 40 years ago. Furthermore, the Chinese were only able to achieve those flights with Russian assistance and it is worth remembering, in light of the Moon as a goal in the new space race, that the Russians failed.

Section IV

Why Cooperation Won't Work and Why a New Space Race is Looming

China has embarked on an ambitious space program designed to compete with the US in both the civil and military arenas of space exploration and space utilization. Concerns regarding China's military intentions and its ambitions to land taikonauts on the Moon have led some to question whether the US should cooperate with China. Others have argued that any Sino–US cooperation is out of the question, citing concerns of technology leaks or inadvertent assistance, possibly leading to China becoming a more formidable space power.

Given the financial burdens that a space race would impose, it would seem to be in the interests of both the US and China to consider opportunities for cooperation. Such a partnership would ensure that the space infrastructure remains intact for the international community. However, given the extremely limited transparency between the two countries and the technological lead maintained by the US, any incentive to cooperate is unlikely.

9

The case for and against collaboration with China

Space has always been a venue for partnerships and competition, whether the "handshake-in-space" Apollo–Soyuz Test Project (ASTP) in 1975, or the US space race with the Soviets in the late 1950s and early 1960s. Although the space age dawned in competitive mode, today's political and funding realities have shifted the balance more towards cooperation, although, as we shall see, this trend may soon be reversed.

International cooperation is perhaps no better exemplified than by the case of the International Space Station (ISS), with its 16 partner nations. Another opportunity for international cooperation is the Vision for Space Exploration (VSE), although, since NASA has already sacrificed its image as a technology innovator to pursue exploration, it is understandable that it does not want to be further constrained by foreign policy requirements. However, exploration, especially on the scale envisioned by the VSE, demands leadership, which in turn is dependent on foreign policy considerations. While it would be a shame to waste such a rare opportunity for exploration and cooperation, in reality, the VSE must compete in a difficult budget environment. This demand for funds may cannibalize both the ISS and NASA's science programs, making it difficult to realize some, if not all, the goals of the VSE. The environment for international and multilateral space cooperation is further complicated by the complex civil space situation. Whereas, in the late 1950s and early 1960s, there were only two players in space, there are currently half a dozen nations (China, France, Great Britain, Russia, and the US) with full space capabilities, and many more, such as Israel, Ukraine, and Brazil, with partial capabilities. Undeniably, the US remains the dominant player, and NASA has a clear mandate to implement the goals of the VSE. However, whether the agency will be able to attract other nations to return to the Moon, in the same way as it persuaded countries to take part in the ISS program, is uncertain. A part of the reason is that the transatlantic ISS partnership has changed, mostly due to the trade limitations associated with the International Traffic in Arms Regulations (ITAR) and the uncertainty surrounding the future of the Space Shuttle and its impact on the ISS. Plans to retire the Shuttle in 2010 will leave the US without a manned spaceflight capability for at least five years and may signal an erosion of NASA's leadership in space. Exacerbating the situation is the realization that, from a budgetary perspective, it is impossible to fully fund a completed ISS, a new transportation system, *and* return to the Moon. One approach that the US may employ to remedy the situation is to negotiate a win–win exit strategy from the ISS with its partners. Such an agreement would include cooperation on transportation systems to reach the Moon – a strategy that would require strengthening waning alliances with ESA and Japan. Another option would be for the US to forge additional partnerships in space with India and China. Such an agreement would permit Indian astronauts and Chinese taikonauts to visit the ISS, or perhaps the Chinese could build a laboratory module that could be attached to a vacant docking port.[1] Chinese scientists and engineers could then develop components of the next generation of spaceships bound for the Moon and Mars. These, and other suggestions for how the two space powers might collaborate in space, are the subject of regular editorials in newspapers, on blogs, and websites. Consider all the time and money that could be saved if only China and the US could cooperate in low Earth orbit (LEO), the pundits suggest. Just think of all the mutual understanding and respect that would result from such a partnership, they say. Undoubtedly, space cooperation projects exist that could deliver worthwhile benefits to both the US and China at acceptable costs and risk. However, anyone who seriously believes that such collaboration could occur is in serious need of a reality check because, for better or for worse, any partnership between Washington and Beijing is unlikely to be realized any time soon.

CHINA'S COLLABORATIVE EFFORTS

China has signed cooperative space agreements with several countries, including Britain, Canada, France, Pakistan, Russia, and Brazil. For example, China has a cooperative agreement with the University of Surrey Space Centre in Great Britain, which markets microsatellites to perform scientific missions such as Earth surveillance. Needless to say, the Sino–Surrey alliance has not received the approval of the US, which is understandably concerned that microsatellite technology could be easily modified for ASAT purposes. Furthermore, the Sino–Surrey association has caused some concerns among politicians in Britain:

> "There is no doubt about this: Surrey has put China into the space weapons business. I am very alarmed. I am particularly concerned because China seems to be right in the middle of nuclear proliferation, passing technology to North Korea, which helps other rogue states such as Iraq and Libya. This may seem like something far away from home. But it directly affects our own national security. This is all happening under the government that promised us ethical foreign policy. What we have got is no foreign policy."
>
> British Shadow Defense Secretary, Iain Duncan Smith (February, 2001)

Despite all its cooperative space agreements, the international cooperation most

Figure 9.1 High above New Zealand, astronauts Robert Curbeam and Christer Fuglesang work to attach a new truss segment to the ISS on December 12th, 2006. Courtesy NASA.

coveted by the Chinese is inclusion in the ISS venture. The ISS (Figure 9.1) is important to Beijing as much for its political aspects as for its technical utility. Chinese participation in the ISS program would not only be a signal to the international community that China had been accepted into the global family of spacefaring nations, but also serve as a seal of legitimacy for the government in Beijing. One reason for China's non-inclusion is that the international consortium of ISS partners are expected to contribute financial support, provide technological expertise, or both and, until very recently, China had neither. Another, more powerful reason for keeping the Chinese out in the cold is Beijing's appalling human rights abuses – a legacy that doesn't fit well with the ISS program that has demonstrated that countries can peacefully work together. However, human rights issues are not the only obstacles to cooperation.

DANGERS OF COLLABORATION

China's ASAT test

Many analysts assert that if the US is serious about maintaining leadership in space, it should engage the Chinese in the ISS program, perhaps inviting them to dock a Shenzhou capsule at the ISS. Although the US has made several attempts in the direction of collaborating with China, the record shows mixed results. For example, in September, 2006, NASA Administrator, Dr Michael Griffin, visited his Chinese counterpart, Laiyan Sun, in China, to investigate the possibility of cooperating with the China National Space Administration (CNSA). NASA's proposal was to allow the cooperation of Chinese scientists in a mission to deliver the large Alpha Magnetic Spectrometer (AMS) to the ISS. No follow-on activities were announced following the trip, although the Chinese issued a proposal for ongoing dialog between NASA and the CNSA that suggested annual exchanges.[1] Any progress that the meeting between Laiyan and Griffin might have generated was quickly forgotten when, on January 11th, 2007, China conducted its first ASAT test.

The prohibitive cost of collaboration

There are space experts who argue that international cooperation is essential in maintaining a space exploration program and, by collaborating with China, the US will surely save time and money in pursuing the VSE's goals. In reality, the US is already locked into partnerships with more than a dozen nations as a part of the ISS program, including most of Europe. Washington has learned from bitter experience that major international projects almost always end up costing more, taking longer, and delivering less than a national program. While many observers have extolled the benefits of US–Russian cooperation during the ISS program, in reality, the venture was a disaster. First, because Russian hardware was years late in delivery, NASA's costs spiraled out of control. Second, the situation was exacerbated by the billions of dollars wasted in redesigning integration hardware. Third, in exchange for just 5% of the financial contribution, Russia was granted 40% of the station's facilities, in addition to making billions of dollars in foreign sales of space hardware! Not surprisingly, from a financial perspective, the US–Russian cooperation experience is one that the Americans will not want to repeat by collaborating with the Chinese.

Diplomacy in orbit has no effect on Earth

One suggestion made by analysts such as Taylor Dinkerman, a spaceflight observer writing for the space policy site *Space Review*, has been for the Americans to engage the Chinese in a space project to generate at least a minimal level of political trust. By pursuing this course of action, analysts hope that by cooperating in space, the political relationship between Washington and Beijing can be changed for the better. Unfortunately, despite what people may think about the supposed benefits that occurred as a result of the US–Russia partnership, "handshakes in space" do not compel world leaders to make peace, no matter how many speeches astronauts and cosmonauts make, extolling the virtues of cooperation. The reason cooperation in space will never help to overthrow old tensions between Washington and Beijing, no matter how many astronauts and taikonauts hug each other in LEO, is that diplomatic progress always comes first.

Technology transfer

China has a long history of acquiring technology by nefarious means. A good example is the launch of China's lunar satellite, Chang'e, which appears to have been adapted from the design of DFH-3, a Chinese communications relay satellite. The DFH-3 was developed in record speed thanks to a large number of Western components used.[2] These components included elements such as the Matra Marconi-manufactured central processor, the infra-red Earth sensor built by Officine Galileo, and parts of the solar panel built by Messerschmitt–Boelkow–Blohm. When the Chinese decided to build the lunar probe, it simply adapted the Western DFH-3 components, enabling them to proceed quickly and reliably.

More recently, the FBI, in conjunction with other US counter-espionage agencies, have tagged more than 100 people and companies allegedly involved in clandestine aerospace technology transfer benefitting China's space program.[3] For example, physicist Shu Quan-Sheng, a naturalized US citizen, was arrested on September 24th, 2008, on charges of illegally exporting space launch technical data and services, in addition to offering bribes to Chinese officials concerning the Long March (LM)-5. Shu, a president of a NASA subcontractor, provided technical assistance and foreign technology acquisition expertise to several of China's government entities involved in the design and development of the LM-5 space launch facility, an activity that the US alleges began in 2008.[4]

In another recent case, US citizen, Ping Cheng, and Singaporeans, Kok Tong Lim and Jian Wei Ding, were charged with conspiracy to violate export administration regulations by attempting to illegally export high-modulus carbon fiber to China. The material, known as Toray M40 and Toray M60, is a corrosion-resistant material used for electromagnetic shielding in rockets and spacecraft.

BARRIERS TO COLLABORATION

Moral compromise

China is widely criticized for its abysmal record on human rights and non-democratic governance. Sadly, the appalling human rights tragedy unfolding every day in China is sidestepped when international cooperation is mentioned, so it is worth providing a reminder.

Human rights violations in China remain systematic and widespread – a situation perpetuated by a government that continues to maintain political control over a legal system in which no one is held accountable. Consequently, abuses such as arbitrary detention, torture, and restrictions of freedom routinely go unchecked. For example, China continues to detain people for exercising their rights to freedom of association and freedom of expression, such as the right to impart and receive information. Persons who exercise these basic rights are regularly detained without charge or trial and deprived of access to legal counsel. The widespread practice of "verdict first, trial second" is still endemic in China's judicial system that lists 60 crimes for which the death penalty can be imposed and, according to Amnesty International, kills 22 prisoners a day. In keeping with its penchant for perpetrating violent acts, China continues to torture its prisoners, and, despite Beijing being a signatory of the UN Convention Against Torture, the government has not implemented measures to reduce the practice. Worse still is the situation in Tibet, where hundreds of Tibetans have been incarcerated for peacefully expressing political beliefs and where Tibetan women are routinely raped, tortured, assaulted and abused.

The previous human rights synopsis is merely the tip of the iceberg of a repressive authoritarian government that suppresses dissent with brutal effectiveness. Any collaboration would inevitably improve the moral standing of Chinese leaders and would therefore require such a moral compromise that would simply be viewed by Western nations as unacceptable. It just isn't going to happen any time soon.

Lack of transparency

Transparency refers to a condition of openness, allowing nations to signal their intentions and capabilities by obtaining or exchanging information on items or activities of interest to the parties involved. Transparency permits international counterparts to increase their confidence about whether an activity is taking place and also provides warning of suspicious behavior – a particularly important consideration for any nation deliberating on doing business with Beijing. But transparency isn't just about sharing perceptions about risks and threats. It requires several important steps, including exchanges between laboratories, information concerning space budgets, operations, research and development programs, and agency-to-agency contacts. Ultimately, transparency requires each counterpart to declare all activities. Such an agreement enables each nation to engage in reciprocal and observable actions that in turn signal a commitment to enforcing predictable rules of behavior.

Transparency is a feature notably absent from China's secrecy-bound space program – a situation exacerbated by the control by the People's Liberation Army (PLA) of virtually all Chinese space development. Such control is clearly a counterproductive factor in any potential agreement with international counterparts. However, even if the PLA wasn't involved, neither Washington nor Beijing believes it confronts a common problem in space that demands mutual collaboration. Furthermore, even if Washington and Beijing investigated the possibility of cooperation *and* engaged in measures to build transparency into their respective space programs, such an attempt would be futile given the disparity in the technological capabilities between the two countries.

Other transparency barriers to collaboration include the obsessive culture of secrecy surrounding the Chinese space program and the reticence of Beijing to reveal just how technologically mature their space hardware is. This reluctance towards efforts in transparency and the insular nature of China's security apparatus have resulted in US efforts to encourage greater bilateral exchanges failing miserably. Furthermore, as long as the US maintains its tremendous technological lead and overwhelming reliance on satellites for military operations and commerce, and as long as China continues to seek parity, the incentives for information exchange will remain slim to non-existent.

ALTERNATIVE FUTURES: COOPERATING WITH CHINA

Unpredicted outcomes

On certain diplomatic levels, the Sino–US relationship is similar to the one that existed between the Soviet Union and the US more than four decades ago. In 1962, at the height of the Cold War, few would have predicted the possibility of a joint spaceflight, but just 10 years later, a bilateral agreement led to the docking of Soyuz and Apollo spacecrafts. While the Apollo–Soyuz Test Program (ASTP) was undoubtedly a significant political and historical event, many analysts still harbored fears about the Soviet Union's ultimate intentions, even after astronaut Thomas Stafford and cosmonaut Alexei Leonov shook hands on July 17th, 1975. Given these suspicions, few would have believed the unprecedented level of cooperation that took place during the Shuttle–Mir era between 1994 and 1998. During this timeframe, American astronauts spent nearly 1,000 days living in orbit with Russian cosmonauts onboard the Russian space station, Mir (Figure 9.2). The Shuttle–Mir program, which witnessed 10 dockings of the Space Shuttle with Mir, not only prepared the way for the ISS, but began an era of cooperation seldom seen in human history. Less than a decade after the end of the Shuttle–Mir program, it was the Russian Soyuz capsule that assured access to space for NASA astronauts following the *Columbia* accident in 2003. It was an outcome few could have predicted.

The point is that it is impossible to predict the future, just as it is impossible to know if or how Sino–US relations might develop. The Soviet–US lesson has taught us that despite fears about the Soviet Union's intentions, informed decisions were made about how the Soviet Union and the US might cooperate in space. These decisions ultimately resulted in a productive international partnership that served to build confidence between the two nations and advanced space exploration. How such a level of cooperation and agreement may be achieved between China and the US is as difficult to predict as the ASTP and the Shuttle–Mir program, but there are some policies and guidelines that, if followed, may enable such collaboration to be realized.

Avoiding a descent into space warfare

While the idea of collaborating militarily with the Chinese is a non-starter, the notion of security collaboration has been suggested as a means of moving towards a common interest. One low-level route to collaboration would be to open treaty negotiations with China on the subject of the military use of space. Another option discussed in the left-leaning sections of the blogosphere is to establish rules of the

Figure 9.2 Space Shuttle docked with Mir space station. Courtesy NASA.

road for space, akin to the code of conduct that exists at sea. Such an agreement would create special caution and safety areas around satellites, provide notification measures, and restrict actions such as ASAT tests.

Space cooperation mechanisms

Given Beijing's history of illegal technology acquisition, the US should not assist China in the research and development of its space programs. The US should only be willing to work with China on certain programs when the Chinese are able to do so at their own pace of development and using their own technology. At the same time, the US should not actively obstruct China's civil space program. Instead, it should adopt a passive stance towards China's development, while also preventing the transfer of any technology with military application that might assist China in achieving military objectives in space.

COLLABORATION REALITY

Sidelining China

The blogosphere seems to be awash with animated statements by newspaper editorial writers and space analysts speculating about the wonderful prospects of Sino–US collaboration. These pundits are in need of a reality check. The US already collaborates with several nations as part of the ISS program. Leaving China on the sidelines causes increased anxiety among the Chinese, who are desperate to join the rest of the world in the space game. Such anxiety to obtain recognition makes it possible for the US to drive a hard bargain if it does eventually decide to engage in talks aimed at collaboration. Meanwhile, the US is happy to pursue its doctrine of space dominance and VSE in a race in which it is setting the benchmark for the competition.

US dominance in space

Ultimately, while arguments can be made for the benefits of cooperation, in reality, pursuing this path would require both the US and China to share resources and technology – a step neither is willing to take, regardless of the potential benefits. Undoubtedly, one of the most important security challenges in the next decade will be how the US deals with China, but it is unlikely that the option of cooperation will be on the table. Some of the reasons why the US will not entertain the notion of collaboration have been discussed in this chapter. Perhaps a more powerful reason is the nature of the national security relationship between Beijing and Washington – a dynamic reminiscent of the US–Soviet relationship in the late 1950s and early 1960s. Back then, the US maintained the high ground in nuclear power, believing that although the Soviets were making progress, the US still had an unmatched ability to decimate the Soviet Union with strategic airpower. After the Sputnik shock, the US had to recalibrate, as evidenced by President Eisenhower's broad educational effort to reassert American leadership in space while raising the public's understanding of the global security situation. The difference this time around is that there will be no Sputnik shock and, with US superiority in space all but assured, there is no incentive for Washington to seek common ground with the Chinese. While the potential clash of interests may not yet be sufficiently severe to be visible to casual observers, the course would appear to be set towards greater *competition* rather than collaboration.

REFERENCES

1. Martel, W.C.; Yoshihara, T. "Averting a Sino–U.S. Space Race", *Washington Quarterly* (Autumn, 2003), 19–35.
2. "China Offers 4-Point Proposal to Boost Sino–U.S. Space Co-Op", The Chinese Government's Official Web Portal, www.gov.cn (September 25, 2006).
3. DeGraffenreid, K.E.; Cox, C. *Report of the Select Committee on U.S. National Security and Military/Commercial Concerns with the People's Republic of China (The Cox Report)*. United States Congress (May 25, 1999).
4. Covault, C.; Ott, J. Caught in the Net: Justice Dept. Implicates Iran, China Contacts in Tech-Transfer Violations. *Aviation Week and Space Technology*, November 3, 34–35 (2008).

10

The imminent space race

> "We dare not tempt them with weakness. For only when our arms are sufficient beyond doubt can we be certain beyond doubt that they will never be employed."
>
> John Fitzgerald Kennedy (1917–1963), Inaugural Address, Washington, DC (January 20, 1961)

The US and China have strong incentives to avoid a new space race, regardless of whether the prize is landing on the Moon or achieving military space dominance.

Militarily, offense has a major advantage over defense, and the US is taking steps to reduce its vulnerability through defensive counterspace techniques. Nevertheless, despite increasing hardening and maneuverability of its satellites, the US is unlikely to achieve more than a modest reduction in space vulnerability. Eventually, the US will confront a situation in which its satellites become increasingly exposed to sophisticated attack – circumstances that will be exacerbated as China's space technology advances.

Meanwhile, in the civil space arena, the Chinese will continue their ambitious space plans while the Space Shuttle is grounded, with the result that the world may perceive China as having overtaken the US in manned space activity. In reality, Chinese technology will not have outpaced US space technology, but Beijing's sustained political commitment will. In this final chapter, these circumstances are considered and the question of whether a new space race is imminent is addressed.

THE NEW MANNED SPACE RACE

Arthur C. Clarke's famous novel, *2010: Odyssey Two*, features the *Tsien Hsue Shen*, a Chinese manned interplanetary spaceship travelling to Jupiter and its moon, Europa, ahead of the Americans and the Russians (Panel 10.1). As in many of Arthur C. Clarke's novels, the fictional voyage of the *Tsien Hsue Shen* may prove to be prophetic, given China's ambitious space plans.

Panel 10.1. China's pivotal role in Arthur C. Clarke's novel

In his 1982 science fiction novel, *2010: Odyssey Two*, the sequel to the spectacularly successful *2001: A Space Odyssey*, Arthur C. Clarke envisioned China entering a space race and taking the lead, ahead of the US and the Soviet Union. *2010* takes place nine years after the *Discovery*'s disastrous mission to Jupiter. To solve the mystery of what happened to *Discovery*, a joint American–Soviet rescue mission is launched to Jupiter. Much to the chagrin of the US and the Soviet Union, the *Tsien Hsue Shen* arrives at Jupiter first, although disaster eventually befalls the Chinese when they are attacked by an indigenous life-form on Europa.

How China's manned space program may fuel a new space race

Acquiring soft power

Manned spaceflight as an expression of leadership is a commodity as important to security as any missile interceptor or kinetic kill vehicle (KKV). Furthermore, leadership acquired through the successes of a manned spaceflight program is a potent soft power tool. While Washington has long since recognized the potential of manned spaceflight as the basis for soft power influence, it is only recently that Beijing has realized that it, too, should exploit human spaceflight for similar reasons. Undoubtedly, the competition to acquire soft power through a successful manned spaceflight program will be one of the driving forces behind the new space race.

Maintaining leadership in space

Thanks to its high-profile manned space missions, much of the world perceives China as catching up with the space capabilities of the US. In reality, nothing could be further from the truth but, as China continues to accelerate its manned space program, the two nations may eventually approach a critical juncture that will decide whether the US will continue to be considered as the leader in human spaceflight. However, it is highly unlikely the US will abrogate its leadership role in human spaceflight, since this would have strategic consequences beyond the space realm. Equally, the Chinese, bolstered by the media coverage of their successful manned missions, will be determined to maintain their sustained effort and to see their goal of leadership in space through to a successful conclusion.

Predicting the future

Speculation runs rampant on the subject of what China's future manned space activities may be. Analysts disagree on the nature of China's intentions regarding dual-use technology, and opinions differ on the question of how much of Beijing's space program is a reaction to Washington and how much is based on China's own politico-strategic ambitions. While Beijing has been relatively open with its coverage of its Shenzhou launches, China's obsession with concealment is pathological, rooted deep in secrecy laws – a part of the Communist regime's attempt to control information. This culture of secrecy has made speculation a primary source of information, not just in the US, but among analysts everywhere. Consequently, deciphering Beijing's intent is often based on a fusion of information gleaned from official and unofficial sources, data mined from poorly translated technical journals, "intent analysis", and lone blogger sites. Exacerbating the dissemination of information are the techno-political and cultural constraints under which China's space officials operate. Office phones are rarely answered, fax machines don't work, voice mail is almost unheard of, e-mails usually bounce back, and, in the rare cases in which officials *do* present information, they usually stick to officially approved scripts. Occasionally, CASC officials appear at conferences held by associations such as the International Association for the Advancement of Space Safety (IASS) and the National Space Society, where Chinese space officials have welcomed international cooperation on Chinese space programs. Unfortunately, such meetings have been the exception rather than the rule.

China's near-future manned spaceflight activities

In keeping with its progression in the arena of manned spaceflight, China's official plans will probably be incremental, cautious, and, above all, ambitious. In the immediate future, Shenzhou missions 8, 9, and 10 are planned, the end result of which will be an orbiting space laboratory as early as 2011. Although space officials and politicians are mindful of the consequences of an accident befalling the space program, perseverance is recognized as a quality required by all manned spacefaring nations and, in the minds of the Chinese especially, is seen as part of the heroic struggle required to explore space.

On the subject of lunar exploration, China has already launched Chang'e, an unmanned probe to the Moon. Chang'e is the first of a series of three missions. The second mission, which will take place in the 2007–2012 timeframe, will land a rover on the Moon, while the third mission will be a sample-return mission that will take place between 2012 and 2017. The three missions, estimated to cost $1.73 billion, are simply coded "Circle", "Land", and "Return", and, collectively, comprise the "lunar decade" of exploration announced in July, 2006. The lunar decade program is intended as a precursor to a manned mission to the Moon, which will commence by sending a manned mission to orbit the Moon in the 2020–2025 timeframe. Once the Chinese version of *Apollo 8* has succeeded, the Chinese will embark upon the task of building a Moon base. However, these dates, as with almost every piece of information generated by the Chinese, may be misleading. Shortly after NASA announced the goals of the Vision of Space Exploration (VSE), which included returning astronauts to the lunar surface by 2018, Ouyang Ziyuan, head of China's lunar project, asserted that China would make a manned lunar landing in 2017. Chinese officials are also adamant that their manned missions to the Moon will be to build infrastructure for sustained use, rather than a "flags and footprints" mission. If this is true, it may be the Americans who win the race to the Moon (again!), as it is unlikely the Chinese will have mastered the complexities of delivering large payloads from lunar orbit to the surface of the Moon by 2017. Another potential time constraint is the time it takes China to launch a mission. Based on the 11 years it took them from the approval of the first manned flight to actually executing it, the time required to realize a manned lunar mission may be in the order of 10–12 years or more.

Planning for the new space race

The Chinese have already embarked upon the initial steps of their lunar program with the construction of a satellite launch base in Hainan Province that will become operational before 2010. Analysts predict that construction of the launch base will increase the thrust of the Long March (LM) rockets by 10%. However, the new launch base is just one small part of the infrastructure, hardware, and techniques required before a serious attempt can be made to land taikonauts on the Moon. In addition to the new base, the Chinese must develop super-thrust, multiple ignition carrier rockets to carry the taikonauts all the way to the Moon and back. This feat will be achieved by the development of the LM-5, due to be operational by 2014, depending on which source you believe. Among the many other requirements are the development of airlock modules and upgraded space suits capable of functioning in the large temperature variations encountered on the lunar surface.

The Chinese are hoping these goals can be realized sooner rather than later in an attempt to upstage the Americans in the race to the Moon. Such an achievement would not only be a techno-nationalist public relations coup, but would also fit with the Chinese strategic concept of *shi*, which uses deception and intelligence to create opportunities for surprise.

Space race inevitability

Manned spaceflight has been the great human and technological achievement of the past 50 years. By launching astronauts and taikonauts into orbit, the US and China have stirred the imagination of the world while expanding and redefining human experience. In the first few decades of the 21st century, manned spaceflight will continue to generate public enthusiasm and to embody the human drama of exploration. However, the distrust and lack of cooperation between Washington and

Beijing suggest human spaceflight may be overshadowed by a competition in which the two main players seek to display their technological prowess and bolster national prestige. Since both countries recognize that space can provide one country with advantages, or at least avoid disadvantages, compared to the other, such a competition would seem inevitable. For its part, the US understands that the manned space arena can no longer be regarded as their backyard – a fact that is perhaps even more relevant militarily.

THE NEW ARMS RACE IN SPACE

China's military space plans

China's space program has rightly been described as "a mystery within a maze".[1] Although some elements of the program are visible, such as the hardware described in this book, even these components are rarely accessible to close scrutiny. Other dimensions of the program, such as its warfare doctrine, are still subject to the pervasive secrecy and compartmentalization that persist throughout China's military–industrial complex. Given China's obsessive suppression of information, analyzing their military space capabilities is a daunting task, even for a seasoned space analyst. The information presented in this book, therefore, represents merely an overview of a space weaponization program that seeks to compete with the US. However, further evidence for why China and the US may be on the cusp of a new Cold War in space may be found by examining how each nation's dependency on space assets affects international security.

The soft kill doctrine

Beijing's plans for escalating its military space capabilities are even less clear than the goals for its manned space program. However, in common with India, Pakistan, Russia, and the US, China's war-fighting doctrine is moving away from comprehensive war with large platforms to smaller wars utilizing high-tech capabilities. Having observed the US fight its wars in Iraq, the Chinese know that to conduct a technologically sophisticated war, it is necessary to become a space power. By acquiring the capability to control ground stations and interfere with information systems, China believes it can limit the capabilities of its opponents through "soft kill" activities. However, the actions of China in its expansion of military space capabilities are a concern to the US, which believes space-based Chinese hardware could be used in a conflict against Taiwan. The Chinese, on the other hand, believe the deployment of American space-based weapons systems could be used in a preemptive strike against Beijing. These fears were fuelled by *Schriever I*, the first space wargame held in the US in 2001 (Panel 10.2).

Panel 10.2. *Schriever* wargames

The *Schriever* wargames series began in 2001 as a means evaluate space-centric issues and to provide the unique opportunity to investigate future space systems. The wargames also enabled military officers to evaluate ways of ensuring the survivability of space systems in the event of a conflict.

The first *Schriever* wargame heightened concerns among some nations that the US Air Force was moving aggressively towards turning space into a combat medium. Following the 2001 exercise, for example, Chinese Foreign Ministry spokesman, Sun Yuxi, was quoted as saying "China expresses its deep concern over the large-scale simulated space confrontation exercise conducted by the U.S. military".

Additionally, the *Schriever* wargames are intended to build and sustain a cadre of space professionals fluent in space issues and focused on the needs of joint warfighters. The wargames series has continued to expand in both size and complexity in proportion to the growing security requirements of an increasingly interdependent world. This environment dictates the need for a space-specific wargame, although space now plays a role in every major wargame. For example, *Schriever 3*, set in the year 2020, explored ways to cope with the loss of US satellites at the tactical level. As in previous wargames, *Schriever 3* featured so-called space control operations, including the denial of satellite capabilities to US adversaries – a situation that the US may face in a conflict over Taiwan.

Schriever I and the Taiwan issue

On January 26th, 2001, the US Space Warfare Center (SWC) concluded *Schriever I*, set in 2017. Although most of the details of the wargame scenario are classified, it is known that the opponents of the US forces threatened a small island about the same size and location as Taiwan. Unsurprisingly, since *Schriever I*, China's politico-militaristic literature on space-based warfare has expressed concern on the subject of US space ambitions and intentions. Of most concern to Beijing is the knowledge that space-based military assets assure the US of force enhancement capabilities that could prevent China from entering into a conflict with Taiwan. This possibility is especially alarming to the Chinese, whose doctrine is to take Taiwan by a surprise, multi-pronged attack followed by bombing, a naval blockade, occupation of air bases, and deployment of special operations forces. The objective of such an assault would be to subjugate Taiwan and secure a *fait accompli* before the US could intervene. China's concern is that US space dominance could greatly reduce the chances of success for this "rapid war, rapid resolution" strategy. One way to neutralize the US advantage would be simply to employ an anti-satellite (ASAT) program. However, due to the limitations of China's tracking abilities, any such program would be limited and be incapable of ensuring an asymmetric response. Also, given the limitations of China's military budget, which is eight times smaller than the US's, it is impossible for the Chinese to even come close to matching US capabilities in space, which only leaves an asymmetric option such as an ASAT program. Ultimately, the Taiwan issue will not only increasingly define Beijing's space warfare strategy, but will also become the fulcrum of the American strategy. More significantly, pursuing these strategies places China and the US on a collision course in space and will result in an inevitable arms race in space.

China's threat to the US

First, China's space program enables Beijing to expand its national power – a strategic objective critical to recovering the greatness that China benefitted from for much of the last millennium. In this regard, China's participation in the new space race is as much a battle with its own demons as it is a quest for prestige. To recover its former greatness, China must not only sustain high levels of economic growth and preserve national stability, but must also defuse external threats to its national security. As described in Chapters 1 and 4, China's space program helps to advance these goals simultaneously. In short, space technologies have now become critical to the successful conduct of China's military operations, while simultaneously procuring symbolic benefits via its manned space program.

Second, China's space program is comprehensive, integrated, and focused. As evidenced by its extensive military and civil hardware, China is now a major spacefaring nation capable of pursuing the entire spectrum of space activities. Unlike the US, which funds a civilian *and* a military space program, no significant divide exists between China's civilian and military programs. Additionally, the focus of China's space program, while undoubtedly marked by deliberation and purpose, is nevertheless characterized by a "buy, copy, or steal" approach, aimed at competing with the US technologically.

Third, Chinese doctrine is undeniably offensive. Due to the prospective future of arms in space, Beijing is forced to develop the entire spectrum of capabilities required to exploit space. It must also prepare to deny space to potential adversaries such as the US, who are capable of utilizing their military advantage to neutralize Chinese space assets. Furthermore, given the pivotal significance of information domination in producing victory in war, China must prepare for future space conflicts by fully integrating its space assets into its military operations.

Fourth, in common with the US, China has made considerable investments in developing counterspace capabilities. These capabilities range from space object surveillance and identification systems to directed energy weapons, and from co-orbital satellites to kinetic attack technologies. Furthermore, these counterspace programs have persisted even after China's ASAT test – an event that conclusively demonstrated China's true intentions in low Earth orbit (LEO).

> "Assuring the security of space capabilities becomes more challenging as technology proliferates and access to it by potentially hostile entities becomes easier. The loss of space systems that support military operations or collect intelligence would dramatically affect the way American forces could fight, likely raising the cost of lives and property and making the outcome less secure. American space systems, including the ground, communication and space segments, need to be defended in order to ensure their viability."[2]

The points addressed here may have cumulative consequences for US national security. Not only will China be able to increasingly disrupt the US's ability to maintain space situational awareness (SSA), but the expansion of China's counter-space capabilities make the discharge of US security obligations much more cumbersome. This means that in the event of a conflict over Taiwan, the US space-based power projection operation will be confronted by China's space denial capabilities. Such a situation will ensure a more extensive contest that may be to the US's disadvantage. Perhaps more importantly, the high frontier will increasingly become weaponized as the US moves away from vulnerable space platforms and towards more robust and maneuverable satellites designed to survive counterspace attacks.

The case against an arms race in space

It has been argued that engaging Beijing in an arms race in space would initiate a spending extravaganza in China, with the result that the Chinese would bankrupt themselves in an attempt to keep up with the US. While there is some substance to this analogy, it can be argued China has no reason to engage in a race to weaponize space because no matter how much money is spent, it could never hope to keep up with or outpace the US. Additionally, severe structural weaknesses underlie China's economic growth. For example, it can be argued that many aspects of China's economic and military progress are exaggerated. A case can also be made that China has a greater commercial interest in the rest of the world than it does an ideological interest. Finally, a strong argument can be made that China has an interest in a healthy US, which can purchase its products and borrow its money. Put simply, China does not want to *bury* the Americans; it wants to *buy* the Americans, and this cannot happen if hostilities occur.

The case for an arms race in space

Beijing's ASAT test in January, 2007, revealed the darker side to China's intentions in space. The test convinced many in the US, most of whom didn't need much persuading, that China's space efforts pose a threat not just to American prestige, but to national security also. The hawks waving the red flags on the subject of Chinese military space capabilities supported their claims by citing official Defense

Department documents, suggesting that China is pursuing the development of all sorts of ASAT weapons, ranging from basic KKVs to stealthy parasitic microsatellites. The latter weapon would be launched into orbit, where it would rendezvous and attach itself to an American satellite. Then, it would lie in wait until it was required, before blowing itself up, thereby killing its target in an event reminiscent of a scene from a Tom Clancey novel (Panel 10.3)!

Panel 10.3. Dubious sources

Some of the information printed in such publications as the Pentagon's annual report, *Military Power of the People's Republic of China*, is based on dubious sources. However, this doesn't prevent the media from releasing the information with the same surprise as the sheep that is flustered to discover the sun rising in the East each morning! Unfortunately, the sometimes dubious claims made by the Pentagon drive the perceptions of American analysts and intelligence officers. In reality, some of what is reported by the Pentagon should be taken with more than a few grains of salt. A good example of questionable intelligence generated by the Pentagon is the "Tom Clancey" parasitic microsatellite, the development of which appeared in the 2003 and 2004 editions of the Defense Department reports to Congress. A checking of facts revealed that the microsatellite was actually the creation of the fevered imagination of a Chinese blogger, who had a history of publishing details of far-fetched weapons.

Dubious sources notwithstanding, China's ASAT demonstration indicated to the world that it could, if it so chose, destroy US satellites in LEO. Equally, the US, with their demonstration of a modified missile-defense interceptor that destroyed a US satellite, demonstrated that it, too, could destroy LEO satellites. Having crossed the ASAT Rubicon, and with the arrival of the Obama Administration, China and the US now face fundamental choices concerning the deployment and use of ASAT capabilities. While the US possesses a far more adept ASAT capability than the Chinese, Washington is also more dependent on space assets than Beijing, and therefore has more to lose in a space war. Furthermore, China's relatively small dependence on space, which compromises its military capability, may actually confer a potential relative near-term offensive advantage, since China has the ability to destroy more US space assets than vice versa. It is an asymmetrical advantage that may persist for a while. Although the US will undoubtedly attempt to neutralize China's asymmetrical advantage by devising ways of inflicting more damage upon China's LEO assets, the Chinese, despite limited technology and financial resources, can easily counter the increased US threat by simply deploying an ASAT fleet. The deployment of fleets of ASATs similar to the one tested in 2007, armed with microwaves and lasers, could significantly reduce the effectiveness of US fighting forces. Simply by destroying half a dozen satellites, the Chinese would create a LEO debris field travelling at 27,000 km/h with the destructive potential to devastate dozens of US satellites, thereby significantly weakening US forces.

The momentum of US space weaponization is another issue increasing the likelihood of a space-borne arms race. In January, 2001, a congressionally mandated space commission called for the President to have the option to deploy weapons in space to deter threats, and, if necessary, to defend against attacks on US interests.[3] This request was followed in 2002 by the US withdrawal from the Anti-Ballistic Missile (ABM) Treaty. Withdrawal from the ABM Treaty was followed by the publication of the US Air Force's (USAF's) vision of how counterspace operations could not only help "achieve and maintain space superiority but also permit the freedom to attack as well as the freedom from attack" in space.[3] Since the USAF's announcement, the US has been pursuing a number of military systems that could attack targets in space from Earth, or targets on Earth, from space. To China, the US's deployment of the Ground-Based Midcourse Missile Defense (GMD) system represents a first step towards space weaponization.[4] The most recent test of the GMD occurred at the end of 2008, in an exercise demonstrating exoatmospheric, multiple-kill vehicles (MKVs), airborne lasers, and interceptors. Further tests will develop what the Chinese fear will be a robust, layered missile defense system capable of neutralizing China's fewer than two dozen single-warhead ICBMs capable of reaching the US. Such a system could be used not only to strategically blackmail China, but also to give the US much more freedom to intervene in China's affairs, such as undermining Beijing's efforts to reunify Taiwan.

Financing the new race in space

Just because they are outgunned financially doesn't mean the Chinese aren't capable of engaging the US in a competition to minimize the orbital technology gap. Beijing has already demonstrated an ability to use dual-use assets, such as the Long March family of launch vehicles, as force enhancers and has acquired technology capable of being modified for use as ASAT weapons. While these developments may cause a few headaches in the Pentagon, in reality, the military advantage in space is overwhelmingly in favor of the Americans and will be for the foreseeable future.

In terms of a potential race between NASA's and the CNSA's manned space programs, the contest could take a number of routes. One scenario is for members of Congress to pursue arguments that the US is in a space race with China in an effort to generate money for cash-strapped NASA, and, by doing so, advance the US manned space program in a similar manner in which the Soviet Union's activities propelled the Apollo program. Such a scenario, it is argued, would serve another purpose, since the Chinese would dramatically increase spending in an attempt to match the space efforts of the US – a course of action that would ultimately benefit the US. However, an argument in which a Sino–US race is compared to the Soviet–US race is fundamentally flawed, since the Cold War featured competitors who *both* started from similar levels of technology and, more importantly, were both motivated to compete. Since the Chinese know they can never match the US in a race for parity and because the costly war efforts in Iraq and Afghanistan make it unlikely that NASA will be allocated more money, one option for the Chinese is simply to perpetuate the notion of a race by supporting an incremental program. By launching one or two significant missions every year, Beijing could create the perception it is winning the race in space. Alternatively, China could combine its capabilities with other like-minded nations, offering countries the opportunity to be partners in an effort to upstage the Americans in LEO and beyond.

Consequences of an arms race in space

The Chinese and the US both like to appear to attach great importance to the prevention of an arms race in space. Each nation publicly advocates peaceful uses of space, such as remote sensing, meteorology, and disaster reduction. In reality, Beijing and Washington recognize that space is gaining ever greater and strategic value. The rapid development of science and technology has provided the conditions for space to become a platform for warfare – a situation China and the US both understand and neither seems willing to avoid. Tests of weapons in space will exacerbate the already serious problem of space debris, while the deployment of weapons in LEO will disrupt strategic balance and stability on a global scale.

Space weaponization could therefore lead to a race in space that turns LEO into a potential battlefield. While the official stance of China is that it will not conduct a space weapons race with the US, in reality, it is unlikely Beijing will sit idly by and watch its strategic interests being jeopardized without taking the necessary measures. While one option for Beijing would be to reach an arms control agreement to prevent space weaponization, it is likely such an effort will fail. This leaves China with the alternate option of developing space-based countermeasures, such as interceptors and/or asymmetric methods, such as ASAT weapons, to counter critical and vulnerable space-based assets such as tracking satellites.

Dawn of the new space race

The material presented in this book suggests that the US and China may be positioning themselves to pursue a new space race. This race for hegemony in space will be fought with two goals in mind. One will be characterized by seeking space dominance militarily and the other will pursue space leadership by landing humans on the Moon.

The relationship between the US and China will remain a complex one and may ultimately be regarded as the most significant bilateral relationship of the first half of the 21st century. The increasingly important issue of space warfare is a high-stakes one with the potential to destabilize the relationship between Washington and Beijing. While the weaponization of space is sure to spark an arms race in space, such a trajectory is a critical and necessary measure. As leader of the international community, the US finds itself in the unenviable position of making decisions for the good of all nations. On the issue of space weaponization, there appears to be no best option, since, regardless of the choice, there will be those who will benefit and those who will suffer. However, the worst choice is to do nothing. The US has decided that space weapons represent a revolutionary military transformation. While such a step is controversial, the nature of international relations and the lessons of history dictate that such transformations begin with the will of a few acting for the benefit of all. By moving forward with space weaponization, the US is moving forward against the fears of the many and harnessing new technologies for a strategy of cooperative advantage for all. Meanwhile, the Chinese Government, while corrupt and repressive, is not collectively stupid. They learned the lessons of the collapse of the Soviet Union; in a direct arms competition with the US, the US wins. Nevertheless, China will continue to seek status as a global military peer competitor with the US.

It has been argued that more progress was made in perfecting the complexities of manned space flight during the eight years between Kennedy's lunar challenge and the landing of *Apollo 11* than in the three decades since the end of the Apollo program. From this perspective, reintroducing the spur of international competition would seem to be a positive development. Unfortunately, the same cannot be said of the specter of a space weapons race that has the potential to progress into a dangerous competition to develop and deploy counterspace capabilities. Such a competition would impose significant costs and produce few lasting strategic advantages, unless the US can dominate offensively, by destroying China's space assets, and defensively, by protecting its own space assets. The question of whether an unrestrained military competition in space is about to unfold remains an open one. However, in the absence of any efforts to manage the emerging competition, it would be remiss not to prepare for the launch of a new arms race in space.

REFERENCES

1. Johnson-Freese, J. *The Chinese Space Program: A Mystery within a Maze*. Krieger Publishing Company.
2. Report of the Commission to Assess United States National Security Space Management and Organization, pursuant to Public Law 106-65 (known as the Space Commission), Executive Summary, p. xiii, available online at www.defenselink.mil/pubs/space20010111.html (January 11, 2001).
3. US Air Force. "Counterspace Operations", Air Force Doctrine Document 2-2.1 (August 2, 2004).
4. Fu Zhigang. "Concerns and Responses: A Chinese Perspective on NMD/TMD", Consultation on NATO Nuclear Policy, National Missile Defense & Alternative Security Arrangements, Ottawa (September 28–30, 2001).

Appendix I

ORBITS

Geostationary orbits

Geostationary orbits (GTOs) are also known as geosynchronous orbits. Satellites in GTO circle the Earth at the same rate as the Earth spins. The Earth takes 23 hours, 56 minutes, and 4.09 seconds to make one full revolution. Based on Kepler's Laws of Planetary Motion, this requires a satellite in GTO to orbit approximately 35,790 km above the Earth. Satellites in GTO are located near the equator, since, at this latitude, there is a constant force of gravity from all directions. At other latitudes, the bulge at the center of the Earth would pull on the satellite. GTOs permit satellites to observe almost a full hemisphere of the Earth, which is why these satellites are used to study large-scale phenomena such as hurricanes. GTOs are also used for communication satellites. The disadvantage of a GTO is that since these satellites are very far away, they have poor resolution. The other disadvantage is that these satellites have trouble monitoring activities near the poles, which is why satellites are deployed into polar orbits.

Polar orbits

Polar orbits have an inclination near 90°, which allows the satellite to see virtually every part of the Earth as the Earth rotates underneath it. It takes approximately 90 minutes for a polar-orbiting satellite to complete one orbit. These satellites have many uses, such as measuring ozone concentrations in the stratosphere or measuring temperatures in the atmosphere.

Sun synchronous orbits

These orbits allow a satellite to pass over a section of the Earth at the same time of day. Since there are 365 days in a year and 360° in a circle, it means that a satellite orbiting in a sun synchronous orbit (SSO) has to shift its orbit by approximately 1° per day. Satellites in SSO orbit at an altitude of between 700 and 800 km. Due to the Earth's equatorial bulge, additional gravitational forces act on the satellite, causing the satellite's orbit to either proceed or recede. SSOs are used for satellites requiring a constant amount of sunlight. Satellites that take pictures of the Earth would work best with bright sunlight, while satellites that measure long-wave radiation would work best in complete darkness.

Inclined orbits

Inclined orbits have an inclination between 0° (equatorial orbit) and 90° (polar orbit). Inclined orbits may be determined by the region on Earth that is of most interest (i.e. an instrument to study the tropics may be best put on a low-inclination satellite), or by the latitude of the launch site. The orbital altitude of these satellites is usually a few hundred kilometers, so the orbital period is only a few hours. These satellites are not sun synchronous, however, so they will view a place on Earth at varying times.

Appendix II

OUTER SPACE TREATY

Born out of anxiety about the Cold War and excitement about the space age, the Outer Space Treaty bars States Parties to the Treaty from placing nuclear weapons or any other weapons of mass destruction in orbit of Earth, installing them on the Moon or any other celestial body, or otherwise stationing them in outer space. As of January 1st, 2008, 98 States have ratified, and an additional 27 have signed, the Outer Space Treaty. The core legal principle of the Treaty declared that everywhere beyond the atmosphere be *res communis*, an international commons rather akin to the "international waters" of the open oceans on Earth. Given the looming rivalry between China and the US, planting rival Moon bases might be sufficient to cause one of the other powers to renounce the agreement. Fortunately, the Treaty has an easy-to-operate escape hatch: signatory States are free to withdraw from the agreement within one year of giving notice. Unfortunately, American or Chinese withdrawal would reduce the Treaty to irrelevance.

The States Parties to the Outer Space Treaty agreed on the following:

Article I

The exploration and use of outer space, including the moon and other celestial bodies, shall be carried out for the benefit and in the interests of all countries, irrespective of their degree of economic or scientific development, and shall be the province of all mankind.

Outer space, including the moon and other celestial bodies, shall be free for exploration and use by all States without discrimination of any kind, on a basis of equality and in accordance with international law, and there shall be free access to all areas of celestial bodies.

There shall be freedom of scientific investigation in outer space, including the moon and other celestial bodies, and States shall facilitate and encourage international co-operation in such investigation.

Article II

Outer space, including the moon and other celestial bodies, is not subject to national appropriation by claim of sovereignty, by means of use or occupation, or by any other means.

Article III

States Parties to the Treaty shall carry on activities in the exploration and use of outer space, including the moon and other celestial bodies, in accordance with international law, including the Charter of the United Nations, in the interest of maintaining international peace and security and promoting international co-operation and understanding.

Article IV

States Parties to the Treaty undertake not to place in orbit around the earth any objects carrying nuclear weapons or any other kinds of weapons of mass destruction, install such weapons on celestial bodies, or station such weapons in outer space in any other manner.

The moon and other celestial bodies shall be used by all States Parties to the Treaty exclusively for peaceful purposes. The establishment of military bases, installations and fortifications, the testing of any type of weapons and the conduct of military manoeuvres on celestial bodies shall be forbidden. The use of military personnel for scientific research or for any other peaceful purposes shall not be prohibited. The use of any equipment or facility necessary for peaceful exploration of the moon and other celestial bodies shall also not be prohibited.

Article V

States Parties to the Treaty shall regard astronauts as envoys of mankind in outer space and shall render to them all possible assistance in the event of accident, distress, or emergency landing on the territory of another State Party or on the high seas. When astronauts make such a landing, they shall be safely and promptly returned to the State of registry of their space vehicle.

In carrying on activities in outer space and on celestial bodies, the astronauts of one State Party shall render all possible assistance to the astronauts of other States Parties.

States Parties to the Treaty shall immediately inform the other States Parties to the Treaty or the Secretary-General of the United Nations of any phenomena they discover in outer space, including the moon and other celestial bodies, which could constitute a danger to the life or health of astronauts.

Article VI

States Parties to the Treaty shall bear international responsibility for national activities in outer space, including the moon and other celestial bodies, whether such activities are carried on by governmental agencies or by non-governmental entities, and for assuring that national activities are carried out in conformity with the provisions set forth in the present Treaty. The activities of non-governmental entities in outer space, including the moon and other celestial bodies, shall require authorization and continuing supervision by the appropriate State Party to the Treaty. When activities are carried on in outer space, including the moon and other celestial bodies, by an international organization, responsibility for compliance with this Treaty shall be borne both by the international organization and by the States Parties to the Treaty participating in such organization.

Article VII

Each State Party to the Treaty that launches or procures the launching of an object into outer space, including the moon and other celestial bodies, and each State Party from whose territory or facility an object is launched, is internationally liable for damage to another State Party to the Treaty or to its natural or juridical persons by such object or its component parts on the Earth, in air or in outer space, including the moon and other celestial bodies.

Article VIII

A State Party to the Treaty on whose registry an object launched into outer space is carried shall retain jurisdiction and control over such object, and over any personnel thereof, while in outer space or on a celestial body. Ownership of objects launched into outer space, including objects landed or constructed on a celestial body, and of their component parts, is not affected by their presence in outer space or on a celestial body or by their return to the Earth. Such objects or component parts found beyond the limits of the State Party to the Treaty on whose registry they are carried shall be returned to that State Party, which shall, upon request, furnish identifying data prior to their return.

Article IX

In the exploration and use of outer space, including the moon and other celestial bodies, States Parties to the Treaty shall be guided by the principle of co-operation and mutual assistance and shall conduct all their activities in outer space, including the moon and other celestial bodies, with due regard to the corresponding interests of all other States Parties to the Treaty. States Parties to the Treaty shall pursue studies of outer space, including the moon and other celestial bodies, and conduct exploration of them so as to avoid their harmful contamination and also adverse changes in the environment of the Earth resulting from the introduction of extraterrestrial matter and, where necessary, shall adopt appropriate measures for this purpose. If a State Party to the Treaty has reason to believe that an activity or experiment planned by it or its nationals in outer space, including the moon and other celestial bodies, would cause potentially harmful interference with activities of other States Parties in the peaceful exploration and use of outer space, including the moon and other celestial bodies, it shall undertake appropriate international consultations before proceeding with any such activity or experiment. A State Party to the Treaty which has reason to believe that an activity or experiment planned by another State Party in outer space, including the moon and other celestial bodies, would cause potentially harmful interference with activities in the peaceful exploration and use of outer space, including the moon and other celestial bodies, may request consultation concerning the activity or experiment.

Article X

In order to promote international co-operation in the exploration and use of outer space, including the moon and other celestial bodies, in conformity with the purposes of this Treaty, the States Parties to the Treaty shall consider on a basis of equality any requests by other States Parties to the Treaty to be afforded an opportunity to observe the flight of space objects launched by those States. The nature of such an opportunity for observation and the conditions under which it could be afforded shall be determined by agreement between the States concerned.

Article XI

In order to promote international co-operation in the peaceful exploration and use of outer space, States Parties to the Treaty conducting activities in outer space, including the moon and other celestial bodies, agree to inform the Secretary-General of the United Nations as well as the public and the international scientific community, to the greatest extent feasible and practicable, of the nature, conduct, locations and results of such activities. On receiving the said information, the Secretary-General of the United Nations should be prepared to disseminate it immediately and effectively.

Article XII

All stations, installations, equipment and space vehicles on the moon and other celestial bodies shall be open to representatives of other States Parties to the Treaty on a basis of reciprocity. Such representatives shall give reasonable advance notice of a projected visit, in order that appropriate consultations may be held and that maximum precautions may betaken to assure safety and to avoid interference with normal operations in the facility to be visited.

Article XIII

The provisions of this Treaty shall apply to the activities of States Parties to the Treaty in the exploration and use of outer space, including the moon and other celestial bodies, whether such activities are carried on by a single State Party to the Treaty or jointly with other States, including cases where they are carried on within the framework of international intergovernmental organizations.

Any practical questions arising in connection with activities carried on by international intergovernmental organizations in the exploration and use of outer space, including the moon and other celestial bodies, shall be resolved by the States Parties to the Treaty either with the appropriate international organization or with one or more States members of that international organization, which are Parties to this Treaty.

Article XIV

1. This Treaty shall be open to all States for signature. Any State which does not

sign this Treaty before its entry into force in accordance with paragraph 3 of this article may accede to it at anytime.

2. This Treaty shall be subject to ratification by signatory States. Instruments of ratification and instruments of accession shall be deposited with the Governments of the United Kingdom of Great Britain and Northern Ireland, the Union of Soviet Socialist Republics and the United States of America, which are hereby designated the Depositary Governments.
3. This Treaty shall enter into force upon the deposit of instruments of ratification by five Governments including the Governments designated as Depositary Governments under this Treaty.
4. For States whose instruments of ratification or accession are deposited subsequent to the entry into force of this Treaty, it shall enter into force on the date of the deposit of their instruments of ratification or accession.
5. The Depositary Governments shall promptly inform all signatory and acceding States of the date of each signature, the date of deposit of each instrument of ratification of and accession to this Treaty, the date of its entry into force and other notices.
6. This Treaty shall be registered by the Depositary Governments pursuant to Article 102 of the Charter of the United Nations.

Article XV

Any State Party to the Treaty may propose amendments to this Treaty. Amendments shall enter into force for each State Party to the Treaty accepting the amendments upon their acceptance by a majority of the States Parties to the Treaty and thereafter for each remaining State Party to the Treaty on the date of acceptance by it.

Article XVI

Any State Party to the Treaty may give notice of its withdrawal from the Treaty one year after its entry into force by written notification to the Depositary Governments. Such withdrawal shall take effect one year from the date of receipt of this notification.

Article XVII

This Treaty, of which the English, Russian, French, Spanish and Chinese texts are equally authentic, shall be deposited in the archives of the Depositary Governments. Duly certified copies of this Treaty shall be transmitted by the Depositary Governments to the Governments of the signatory and acceding States.

Appendix III

SPACE WEAPON TECHNOLOGY AND PROGRAMS

Aegis Ballistic Missile Defense

Aegis Ballistic Missile Defense is the sea-based portion of the Ballistic Missile Defence (BMD) system. It incorporates the Aegis Weapon System, the Lightweight Exoatmospheric Projectile Intercept, the Standard Missile-3 (SM-3), and the Navy Ballistic Missile systems. Eighteen US Navy Aegis-equipped ships and two Japanese Aegis-equipped destroyers have the capability to engage short to intermediate-range ballistic missile threats and support other BMDS engagements using the Aegis BMD Weapon System and the SM-3.

Common Aero Vehicle/Hypersonic Technology Vehicle

The Common Aero Vehicle (CAV) was originally conceived as an unmanned spacecraft that would travel at five times the speed of sound, carrying munitions or troops from the US to anywhere in the world within two hours. In 2004, the offensive strike part of the project was cancelled and the CAV was renamed the Hypersonic Technology Vehicle (HTV). When operational, the HTV will attain Mach 19, briefly exit the Earth's atmosphere and re-enter, after flying between 30 and 45 km above the planet's surface. During the early part of the flight, the HTV will act like a spacecraft. In the mid-phase of the flight, the HTV will re-enter the atmosphere like the Space Shuttle, and, in the final stage, it will operate like an aircraft.

Defense Support Program

The Defense Support Program (DSP) is a satellite system that warns the US military of missile launches. The 23 satellites in the system use infrared detectors to sense heat from missile plumes, in order to detect missile launches, space launches, and nuclear detonations. DSP satellites have been the space-borne segment of NORAD's Tactical Warning and Attack Assessment System since 1970, but they are to be replaced by the Space Based Infrared System (SBIRS). Typically, DSP satellites are launched into geosynchronous orbit on a Titan IV booster and inertial upper stage combination, although one DSP satellite was launched using the Space Shuttle on mission STS-44 (November 24th, 1991).

Demonstration of Autonomous Rendezvous Technology space vehicle

The Demonstration of Autonomous Rendezvous Technology (DART) space vehicle is designed to demonstrate technologies required for a spacecraft to locate and rendezvous, or maneuver close to other vehicles in space. The Autonomous Rendezvous and Proximity Operations software on DART will test additional algorithms by calculating and executing collision avoidance maneuvers and circumnavigation. As with the XSS-11 microsatellite, the capacity that enables spacecraft to maneuver around others to service them can also allow it to destroy them.

Escort satellites

Escort satellites carry sensors, lightweight missiles, and other weapon systems such as lasers, high-powered microwave systems, and kinetic weapons to defend distant satellites against an anti-satellite attack. They can permanently or temporarily disable large satellites, giving them the ability to carry out anti-satellite (ASAT) attacks.

Exoatmospheric Kill Vehicle

The Exoatmospheric Kill Vehicle (EKV) is the intercept component of the Ground Based Interceptor (GBI), the weapon element of the Ground-based Midcourse Defense system. It is designed to take out hostile ballistic missile targets outside the atmosphere while the missiles are in flight. The EKV has a sensitive, long-range electro-optical infrared seeker enabling it to acquire and track targets, and to discriminate between the intended target re-entry vehicle and other objects. Weighing approximately 64 kg and measuring 120 cm in length and 60 cm in diameter, the EKV is supposed to fly through space at 7,200 km/h and smash into an incoming warhead.

Experimental Spacecraft System-11 (XSS-11)

The XSS-11 is a 100-kg microsatellite that is able to "meet" with other space objects in orbit, and maneuver close to them to inspect them or perform maintenance tasks. However, the XSS-11 could easily be used as an ASAT weapon by either disabling or destroying an adversary's satellite. The XSS-11 could also house a small kinetic-kill vehicle designed to impact into an enemy satellite. With a price-tag of $56 million per spacecraft, the XSS-11 weighs 138 kg and has a design life of between 12 and 18 months.

Force Application and Launch from Continental United States (FALCON)

FALCON was originally a Defense Advanced Research Projects Agency (DARPA)/ Air Force project intended to develop a reusable Hypersonic Cruise Vehicle (HCV), a Common Aero Vehicle (CAV), and a Small Launch Vehicle (SLV). In 2004, FALCON became Falcon, the CAV became the Hypersonic Technology Vehicle (HTV), and, in 2005, Lockheed Martin received $11.7 million for the second phase of the SLV program. FALCON's HCV could lead to an aircraft capable of putting a satellite and humans into space. The project's SLV, for example, should be capable of carrying a 1,000-kg satellite into sun-synchronous orbit. After take-off, a hypersonic craft would use a supersonic turbine engine to reach speeds of Mach 3. At this speed, scramjet engines would take over and use the aircraft's speed to compress air for combustion. Once the vehicle reached its max hypersonic speed, it would then deploy either a separate craft to reach space, as planned for the SLVs, or switch from its air-breathing scramjet engine to some sort of rocket propulsion.

Ground-based Midcourse Defense system

Located in Fort Greeley, Alaska, with a sister-site at Vandenberg Air Force Base in California, the Ground-based Midcourse Defense (GMD) system is the heart of the US missile defense system. It is designed to intercept long-range missiles targeted against the continental US (CONUS). Theoretically, the GMD is dual-use, since it will be capable of shooting down satellites as well as downing intercontinental ballistic missiles. Although the media have reported technical issues associated with the GMD, in reality, the development of the system has been very successful, especially when one considers the complexity of operations and the need to integrate several military components.

Example of GMD Test: September 28th, 2007

On September 28th, 2007, the Missile Defense Agency conducted a successful test of the GMD system, including an intercept of a target missile. The Kodiak Launch Complex (Alaska) launched a long-range ballistic missile target, traveling southward to resemble the trajectory of a North Korean missile. The Early Warning Radar at Beale Air Force Base in California located and tracked the target. Seventeen minutes later, Vandenberg Air Force Base launched an interceptor missile, which released an EKV. The interceptor successfully destroyed the target warhead, marking the seventh successful intercept of the GMD system, and the second time an operationally configured interceptor has been used in the past 13 months. The test integrated the Sea-Based X-Band Radar (SBX) 1 and an Aegis ballistic missile defense ship using its onboard SPY-1 radar to track the target warhead.

Kinetic Energy Interceptor

Kinetic Energy Interceptors (KEIs) are missiles launched into space to destroy enemy missiles by impacting them, rather than by exploding near them. KEIs also have potential applications as ASAT weapons, because the same technology is necessary to destroy incoming missiles and satellites. The relative velocity (closing rate) of a KEI intercept may vary from a low of 1–2 km/second up to a hypervelocity of 8–10 km/second (10 km/second = 36,000 km/h).

Lightweight Exoatmospheric Projectile Program

The LEAP Program is aimed at developing and integrating miniature kinetic energy (hit-to-kill) interceptors to be used in the Aegis BMD system. The LEAP vehicles are intended to intercept enemy missiles within a range of 1,000–2,000 km away, by homing in on the missile and destroying it by direct impact. The LEAP vehicles would also be capable of reaching low-altitude satellites for ASAT attacks.

Miniature Kill Vehicle

The Miniature/Multiple Kill Vehicle (MKV) is an anti-missile interceptor warhead intended to destroy multiple ballistic missile re-entry vehicles and decoys with a single launch. It will consist of multiple small kill vehicles designed to destroy ballistic missile warheads as well as deployed decoys in space by colliding with them during the midcourse stage of the flight. The MKVs are intended to weigh 2–10 kg each and 12 would be launched on a single rocket. Once launched, the rocket's vehicle carrier would use its own sensors to distinguish warheads from decoys and program the MKVs to attack multiple targets. The MKV warhead is being designed to be integrated seamlessly in the Ground-based Midcourse Defense Segment (GMDS).

Multiple Kill Vehicle Payload System

The Multiple Kill Vehicle (MKV) system will be attached to an interceptor. In the event of an enemy launch, a single interceptor equipped with the MKV payload would destroy not only the re-entry vehicle, but also all credible threat objects, including any countermeasures that the enemy deploys to try to spoof US defenses. The system consists of a carrier vehicle with onboard sensors and a number of small, simple kill vehicles that can be independently targeted against objects in a threat cluster. The integrated payload is designed to fit on existing and planned interceptor boosters.

Nanosatellites/Miniature Spacecraft

The term "nanosatellite" or "nanosat" is usually applied to a satellite with a mass of between 1 and 10 kg. One recent nanosat program was the Autonomous Nanosatellite Guardian for Evaluating Local Space (ANGELS). Announced in 2005, ANGELS is designed to provide "localized" space situational awareness and "anomaly characterization" for host satellites in Geosynchronous Orbit (GEO) as well as high-value space asset defensive capabilities. On May 18th, 2004, China successfully launched Nano-satellite I. According to the PRC, the tiny Chinese Nano-satellite was "designed for high-tech experiments".

Near Field Infrared Experiment

The Near Field Infrared Experiment (NFIRE) comprises a satellite with an onboard sensor and Laser Communication Terminal (LCT), as well as two ground-based control centers. It will also carry releasable (and controversial) kill vehicles. The NFIREs sensor is designed to distinguish between a missile and its plume. The primary payload is the Track Sensor Payload (TSP), which will be used to collect the images of the boosting rocket, while the secondary payload is an LCT that will be used to evaluate the utility of laser communications for missile defense applications.

Operationally Responsive Spacelift

The Air Force began the Operationally Responsive Spacelift (ORS) initiative in 2003. The goal of the program is to pave the way for reusable rockets that could be launched at a low cost on short notice. As part of a one-year Analysis of Alternatives study, teams investigated a variety of spaceplanes, air-launched boosters, and fully reusable as well as expendable or partly reusable launch vehicles. The program's payloads include the common aero vehicle (CAV), a munition that can be delivered from or through space, along with counterspace payloads. Another concept under consideration is a hybrid vehicle consisting of a reusable first stage with an expendable upper stage stack. The name given to this next-generation family of hybrid vehicles is Affordable REsponsive Spacelift (ARES). The ARES concept is that of a reusable fly-back booster with expendable upper stages that lends itself to incorporation into a modular family of vehicles, thus permitting support of a range of payload mass insertion needs and a flight rate and fleet size tailored for an optimal balance between responsiveness, affordability, and survivability. A multi-staged system could be in place by 2014, depending on funding.

Orbital Boost Vehicle

The Orbital Boost Vehicle (OBV) is used in the Ground-based Midcourse Defense segment of the US BMD program. It is designed to intercept and destroy long-range enemy missiles while they are in flight (as opposed to Kinetic Energy Interceptors, which are intended to intercept and destroy enemy missiles just after they are launched). The first OBV was launched from Vandenberg Air Force Base, CA, on September 28th, 2007, by Orbital Sciences Corporation.

Sea-based Midcourse Defense system

The Sea-based Midcourse Defense (SMD) system is composed of ship-based missile defense systems intended to intercept incoming missiles above the atmosphere. It incorporates the Standard Missile-3 (SM-3), Aegis Ballistic Missile Defense system, and the Lightweight Exoatmospheric Projectile Intercept program. It also has anti-satellite capabilities.

Small Launch Vehicle

The Small Launch Vehicle is part of the FALCON weapons system. The goal of the program is to develop an affordable space lift capability that can quickly launch small satellites and CAVs/HTVs into orbit.

Space-Based Infrared System

Designed to operate worldwide 24 hours a day, 7 days a week, the Space-Based Infrared System (SBIRS) consists of a constellation of satellites capable of tracking ballistic missiles throughout their course. SBIRS is managed by the SBIRS Program Office at the Space and Missile Systems Center, Los Angeles Air Force Base, California. The SBIRS program addresses critical warfighter needs in the areas of missile warning, missile defense and battle-space characterization.

Space-Based Surveillance System

The Space-Based Surveillance System (SBSS) detects and tracks space objects such as satellites and debris. The Department of Defense intends to use its data to support military operations. The SBSS concept will eventually comprise four satellites in equatorial orbits at an altitude of 1,100 km with the possibility of additional satellites in inclined orbits for polar coverage.

Space Tracking and Surveillance System

The Missile Defense Agency's Space Tracking and Surveillance System (STSS) will be able to detect and track ballistic missiles as well as potential ground-based kinetic energy anti-satellite weapons. In May, 2009, the STSSs Demonstration Satellite Program shipped one of two planned satellites (Space Vehicle-2) to the Eastern Launch Site at Cape Canaveral, Florida. Space Vehicle-2 will be closely followed by its sister vehicle (Space Vehicle-1!) in preparation for a tandem launch planned for August, 2009. Data from STSS satellites will be used by BMDS interceptors to engage ballistic missiles earlier in flight and pick out the warhead of an incoming missile from other nearby objects such as decoys. As technology matures and lessons are learned from the first satellites, more capable satellites will be designed and launched.

Standard Missile-3

The Standard Missile-3 (SM-3) is a ballistic missile that destroys incoming ballistic missiles outside the Earth's atmosphere. It is an integral component of the Aegis Ballistic Missile Defence System.

Standard Missile-3 concept of operations

The SM-3 uses a booster and dual thrust rocket motor for the first and second stages and a steering control section and midcourse missile guidance for maneuvering in the atmosphere. To support the extended range of an exoatmospheric intercept, additional missile thrust is provided in a third stage. The third stage contains a dual pulse rocket motor for the early exoatmospheric phase of flight and a Lightweight Exoatmospheric Projectile (LEAP) Kinetic Warhead (KW) for the intercept phase. Following second stage separation, the first pulse burn of the Third Stage Rocket Motor (TSRM) provides thrust to maintain the missile's trajectory into space. Upon entering space, the third stage coasts, after which the TSRM's attitude control system maneuvers the third stage to eject the nosecone exposing the KW's infrared seeker. If the third stage requires a course correction for an intercept, the rocket motor begins a second pulse burn. Upon completion of the second pulse burn, the infrared seeker is calibrated and the KW ejects. The KW is now on its own and relies on its infrared seeker to acquire the target. Once the target is acquired, the KW uses its own attitude control and guidance system to intercept.

Index

Printing: Mercedes-Druck, Berlin
Binding: Stein+Lehmann, Berlin